TOPOLOGY FROM THE
DIFFERENTIABLE VIEWPOINT

PRINCETON LANDMARKS
IN MATHEMATICS AND PHYSICS

Non-standard Analysis,
by Abraham Robinson

General Theory of Relativity,
by P.A.M. Dirac

Angular Momentum in Quantum Mechanics,
by A. R. Edmonds

Mathematical Foundations of Quantum Mechanics,
by John von Neumann

Introduction to Mathematical Logic,
by Alonzo Church

Convex Analysis,
by R. Tyrrell Rockafellar

Riemannian Geometry,
by Luther Pfahler Eisenhart

The Classical Groups,
by Hermann Weyl

Topology from the Differentiable Viewpoint,
by John Milnor

TOPOLOGY FROM THE DIFFERENTIABLE VIEWPOINT

Revised Edition

John W. Milnor

BASED ON NOTES BY DAVID W. WEAVER

PRINCETON UNIVERSITY PRESS
PRINCETON, NEW JERSEY

Published by Princeton University Press, 41 William Street,
Princeton, New Jersey 08540
In the United Kingdom: Princeton University Press, Chichester, West Sussex

Copyright © 1965 by the Rector and Visitors of the University of Virginia
Reprinted by arrangement with The University Press of Virginia

All Rights Reserved. No part of this book may be reproduced or transmitted in any form or by any means, electronic or mechanical, including photocopying, recording, or by any information storage and retrieval system, without permission in writing from the Publisher.

Library of Congress Cataloging-in-Publication Data
Milnor, John Willard, 1931–
Topology from the differentiable viewpoint / by John W. Milnor ; based on notes by David W. Weaver. (Princeton Landmarks in Mathematics and Physics)
p. cm.
Originally published: Charlottesville : University Press of Virginia, 1965.
Includes bibliographical references and index.
ISBN 0-691-04833-9
1. Differential topology. I. Title.
Qa613.6.M55 1997
514'.72—dc21   97-30986

First printing, in the Princeton Landmarks in Mathematics
and Physics series, 1997

http://pup.princeton.edu

Printed and bound by CPI Group (UK) Ltd, Croydon, CR0 4YY

ISBN-13: 978-0-691-04833-8 (pbk.)

To Heinz Hopf

PREFACE

THESE lectures were delivered at the University of Virginia in December 1963 under the sponsorship of the Page-Barbour Lecture Foundation. They present some topics from the beginnings of topology, centering about L. E. J. Brouwer's definition, in 1912, of the *degree* of a mapping. The methods used, however, are those of differential topology, rather than the combinatorial methods of Brouwer. The concept of *regular value* and the theorem of Sard and Brown, which asserts that every smooth mapping has regular values, play a central role.

To simplify the presentation, all manifolds are taken to be infinitely differentiable and to be explicitly embedded in euclidean space. A small amount of point-set topology and of real variable theory is taken for granted.

I would like here to express my gratitude to David Weaver, whose untimely death has saddened us all. His excellent set of notes made this manuscript possible.

J. W. M.

Princeton, New Jersey
March 1965

CONTENTS

	Preface	vii
1.	Smooth manifolds and smooth maps	1
	Tangent spaces and derivatives	2
	Regular values	7
	The fundamental theorem of algebra	8
2.	The theorem of Sard and Brown	10
	Manifolds with boundary	12
	The Brouwer fixed point theorem	13
3.	Proof of Sard's theorem	16
4.	The degree modulo 2 of a mapping	20
	Smooth homotopy and smooth isotopy	20
5.	Oriented manifolds	26
	The Brouwer degree	27
6.	Vector fields and the Euler number	32
7.	Framed cobordism; the Pontryagin construction	42
	The Hopf theorem	50
8.	Exercises	52
	Appendix: Classifying 1-manifolds	55
	Bibliography	59
	Index	63

TOPOLOGY FROM THE DIFFERENTIABLE VIEWPOINT

§1. SMOOTH MANIFOLDS AND SMOOTH MAPS

FIRST let us explain some of our terms. R^k denotes the k-dimensional euclidean space; thus a point $x \in R^k$ is an k-tuple $x = (x_1, \cdots, x_k)$ of real numbers.

Let $U \subset R^k$ and $V \subset R^l$ be open sets. A mapping f from U to V (written $f : U \to V$) is called *smooth* if all of the partial derivatives $\partial^n f / \partial x_{i_1} \cdots \partial x_{i_n}$ exist and are continuous.

More generally let $X \subset R^k$ and $Y \subset R^l$ be arbitrary subsets of euclidean spaces. A map $f : X \to Y$ is called *smooth* if for each $x \in X$ there exist an open set $U \subset R^k$ containing x and a smooth mapping $F : U \to R^l$ that coincides with f throughout $U \cap X$.

If $f : X \to Y$ and $g : Y \to Z$ are smooth, note that the composition $g \circ f : X \to Z$ is also smooth. The identity map of any set X is automatically smooth.

DEFINITION. A map $f : X \to Y$ is called a *diffeomorphism* if f carries X homeomorphically onto Y and if both f and f^{-1} are smooth.

We can now indicate roughly what *differential topology* is about by saying that it studies those properties of a set $X \subset R^k$ which are invariant under diffeomorphism.

We do not, however, want to look at completely arbitrary sets X. The following definition singles out a particularly attractive and useful class.

DEFINITION. A subset $M \subset R^k$ is called a *smooth manifold* of *dimension* m if each $x \in M$ has a neighborhood $W \cap M$ that is diffeomorphic to an open subset U of the euclidean space R^m.

Any particular diffeomorphism $g : U \to W \cap M$ is called a *parametrization* of the region $W \cap M$. (The inverse diffeomorphism $W \cap M \to U$ is called a system of *coordinates* on $W \cap M$.)

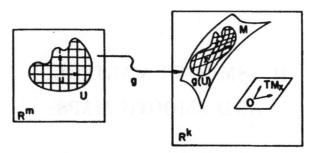

Figure 1. *Parametrization of a region in M*

Sometimes we will need to look at manifolds of dimension zero. By definition, M is a manifold of dimension zero if each $x \in M$ has a neighborhood $W \cap M$ consisting of x alone.

EXAMPLES. The unit sphere S^2, consisting of all $(x, y, z) \in R^3$ with $x^2 + y^2 + z^2 = 1$ is a smooth manifold of dimension 2. In fact the diffeomorphism

$$(x, y) \mapsto (x, y, \sqrt{1 - x^2 - y^2}),$$

for $x^2 + y^2 < 1$, parametrizes the region $z > 0$ of S^2. By interchanging the roles of x, y, z, and changing the signs of the variables, we obtain similar parametrizations of the regions $x > 0$, $y > 0$, $x < 0$, $y < 0$, and $z < 0$. Since these cover S^2, it follows that S^2 is a smooth manifold.

More generally the sphere $S^{n-1} \subset R^n$ consisting of all $(x_1, \cdots, x_n)$ with $\sum x_i^2 = 1$ is a smooth manifold of dimension $n - 1$. For example $S^0 \subset R^1$ is a manifold consisting of just two points.

A somewhat wilder example of a smooth manifold is given by the set of all $(x, y) \in R^2$ with $x \neq 0$ and $y = \sin(1/x)$.

TANGENT SPACES AND DERIVATIVES

To define the notion of *derivative* df_x for a smooth map $f : M \to N$ of smooth manifolds, we first associate with each $x \in M \subset R^k$ a linear subspace $TM_x \subset R^k$ of dimension m called the *tangent space* of M at x. Then df_x will be a linear mapping from TM_x to TN_y, where $y = f(x)$. Elements of the vector space TM_x are called *tangent vectors* to M at x.

Intuitively one thinks of the m-dimensional hyperplane in R^k which best approximates M near x; then TM_x is the hyperplane through the

Tangent spaces

origin that is parallel to this. (Compare Figures 1 and 2.) Similarly one thinks of the nonhomogeneous linear mapping from the tangent hyperplane at x to the tangent hyperplane at y which best approximates f. Translating both hyperplanes to the origin, one obtains df_x.

Before giving the actual definition, we must study the special case of mappings between open sets. For any open set $U \subset R^k$ the *tangent space* TU_x is defined to be the entire vector space R^k. For any smooth map $f: U \to V$ the *derivative*

$$df_x : R^k \to R^l$$

is defined by the formula

$$df_x(h) = \lim_{t \to 0} (f(x + th) - f(x))/t$$

for $x \; \epsilon \; U$, $h \; \epsilon \; R^k$. Clearly $df_x(h)$ is a linear function of h. (In fact df_x is just that linear mapping which corresponds to the $l \times k$ matrix $(\partial f_i/\partial x_j)_x$ of first partial derivatives, evaluated at x.)

Here are two fundamental properties of the derivative operation:

1 (Chain rule). *If $f : U \to V$ and $g : V \to W$ are smooth maps, with $f(x) = y$, then*

$$d(g \circ f)_x = dg_y \circ df_x.$$

In other words, to every commutative triangle

of smooth maps between open subsets of R^k, R^l, R^m there corresponds a commutative triangle of linear maps

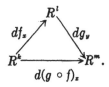

2. *If I is the identity map of U, then dI_x is the identity map of R^k. More generally, if $U \subset U'$ are open sets and*

$$i : U \to U'$$

is the inclusion map, then again di_x is the identity map of R^k.
Note also:

3. If $L : R^k \to R^l$ is a linear mapping, then $dL_x = L$.

As a simple application of the two properties one has the following:

ASSERTION. *If f is a diffeomorphism between open sets $U \subset R^k$ and $V \subset R^l$, then k must equal l, and the linear mapping*
$$df_x : R^k \to R^l$$
must be nonsingular.

PROOF. The composition $f^{-1} \circ f$ is the identity map of U; hence $d(f^{-1})_y \circ df_x$ is the identity map of R^k. Similarly $df_x \circ d(f^{-1})_y$ is the identity map of R^l. Thus df_x has a two-sided inverse, and it follows that $k = l$.

A partial converse to this assertion is valid. Let $f : U \to R^k$ be a smooth map, with U open in R^k.

Inverse Function Theorem. *If the derivative $df_x : R^k \to R^k$ is nonsingular, then f maps any sufficiently small open set U' about x diffeomorphically onto an open set $f(U')$.*

(See Apostol [2, p. 144] or Dieudonné [7, p. 268].)

Note that f may not be one-one in the large, even if every df_x is nonsingular. (An instructive example is provided by the exponential mapping of the complex plane into itself.)

Now let us define the *tangent space* TM_x for an arbitrary smooth manifold $M \subset R^k$. Choose a parametrization
$$g : U \to M \subset R^k$$
of a neighborhood $g(U)$ of x in M, with $g(u) = x$. Here U is an open subset of R^m. Think of g as a mapping from U to R^k, so that the derivative
$$dg_u : R^m \to R^k$$
is defined. Set TM_x equal to the image $dg_u(R^m)$ of dg_u. (Compare Figure 1.)

We must prove that this construction does not depend on the particular choice of parametrization g. Let $h : V \to M \subset R^k$ be another parametrization of a neighborhood $h(V)$ of x in M, and let $v = h^{-1}(x)$. Then $h^{-1} \circ g$ maps some neighborhood U_1 of u diffeomorphically onto a neighborhood V_1 of v. The commutative diagram of smooth maps

Tangent spaces

between open sets

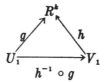

gives rise to a commutative diagram of linear maps

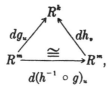

and it follows immediately that

$$\text{Image } (dg_u) = \text{Image } (dh_v).$$

Thus TM_x is well defined.

PROOF THAT TM_x IS AN m-DIMENSIONAL VECTOR SPACE. Since

$$g^{-1} : g(U) \to U$$

is a smooth mapping, we can choose an open set W containing x and a smooth map $F : W \to R^m$ that coincides with g^{-1} on $W \cap g(U)$. Setting $U_0 = g^{-1}(W \cap g(U))$, we have the commutative diagram

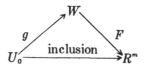

and therefore

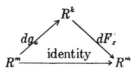

This diagram clearly implies that dg_u has rank m, and hence that its image TM_x has dimension m.

Now consider two smooth manifolds, $M \subset R^k$ and $N \subset R^l$, and a

smooth map
$$f : M \to N$$
with $f(x) = y$. The *derivative*
$$df_x : TM_x \to TN_y$$
is defined as follows. Since f is smooth there exist an open set W containing x and a smooth map
$$F : W \to R^l$$
that coincides with f on $W \cap M$. Define $df_x(v)$ to be equal to $dF_x(v)$ *for all* $v \in TM_x$.

To justify this definition we must prove that $dF_x(v)$ belongs to TN_y and that it does not depend on the particular choice of F.

Choose parametrizations
$$g : U \to M \subset R^k \quad \text{and} \quad h : V \to N \subset R^l$$
for neighborhoods $g(U)$ of x and $h(V)$ of y. Replacing U by a smaller set if necessary, we may assume that $g(U) \subset W$ and that f maps $g(U)$ into $h(V)$. It follows that
$$h^{-1} \circ f \circ g : U \to V$$
is a well-defined smooth mapping.

Consider the commutative diagram

$$\begin{array}{ccc} W & \xrightarrow{F} & R^l \\ {\scriptstyle g}\uparrow & & \uparrow{\scriptstyle h} \\ U & \xrightarrow{h^{-1} \circ f \circ g} & V \end{array}$$

of smooth mappings between open sets. Taking derivatives, we obtain a commutative diagram of linear mappings

$$\begin{array}{ccc} R^k & \xrightarrow{dF_x} & R^l \\ {\scriptstyle dg_u}\uparrow & & \uparrow{\scriptstyle dh_v} \\ R^m & \xrightarrow{d(h^{-1} \circ f \circ g)_u} & R^n \end{array}$$

where $u = g^{-1}(x)$, $v = h^{-1}(y)$.

It follows immediately that dF_x carries $TM_x = \text{Image}(dg_u)$ into $TN_y = \text{Image}(dh_v)$. Furthermore the resulting map df_x does not depend on the particular choice of F, for we can obtain the same linear

Regular values

transformation by going around the bottom of the diagram. That is:
$$df_x = dh_v \circ d(h^{-1} \circ f \circ g)_u \circ (dg_u)^{-1}.$$
This completes the proof that
$$df_x : TM_x \to TN_y$$
is a well-defined linear mapping.

As before, the derivative operation has two fundamental properties:

1. (Chain rule). *If $f : M \to N$ and $g : N \to P$ are smooth, with $f(x) = y$, then*
$$d(g \circ f)_x = dg_y \circ df_x.$$

2. *If I is the identity map of M, then dI_x is the identity map of TM_x. More generally, if $M \subset N$ with inclusion map i, then $TM_x \subset TN_x$ with inclusion map di_x.* (Compare Figure 2.)

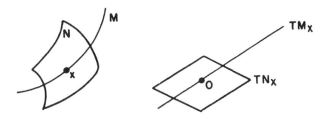

Figure 2. The tangent space of a submanifold

The proofs are straightforward.
As before, these two properties lead to the following:

ASSERTION. *If $f : M \to N$ is a diffeomorphism, then $df_x : TM_x \to TN_y$ is an isomorphism of vector spaces. In particular the dimension of M must be equal to the dimension of N.*

REGULAR VALUES

Let $f : M \to N$ be a smooth map between manifolds of the same dimension.* We say that $x \in M$ is a *regular point* of f if the derivative

---
* This restriction will be removed in §2.

df_x is nonsingular. In this case it follows from the inverse function theorem that f maps a neighborhood of x in M diffeomorphically onto an open set in N. The point $y \in N$ is called a *regular value* if $f^{-1}(y)$ contains only regular points.

If df_x is singular, then x is called a *critical point* of f, and the image $f(x)$ is called a *critical value*. Thus each $y \in N$ is either a critical value or a regular value according as $f^{-1}(y)$ does or does not contain a critical point.

Observe that if M is compact and $y \in N$ is a regular value, then $f^{-1}(y)$ is a finite set (possibly empty). For $f^{-1}(y)$ is in any case compact, being a closed subset of the compact space M; and $f^{-1}(y)$ is discrete, since f is one-one in a neighborhood of each $x \in f^{-1}(y)$.

For a smooth $f : M \to N$, with M compact, and a regular value $y \in N$, we define $\#f^{-1}(y)$ *to be the number of points in* $f^{-1}(y)$. The first observation to be made about $\#f^{-1}(y)$ is that it is locally constant as a function of y (where y ranges only through regular values!). I.e., *there is a neighborhood $V \subset N$ of y such that* $\#f^{-1}(y') = \#f^{-1}(y)$ *for any* $y' \in V$. [Let $x_1, \cdots, x_k$ be the points of $f^{-1}(y)$, and choose pairwise disjoint neighborhoods $U_1, \cdots, U_k$ of these which are mapped diffeomorphically onto neighborhoods $V_1, \cdots, V_k$ in N. We may then take

$$V = V_1 \cap V_2 \cap \cdots \cap V_k - f(M - U_1 - \cdots - U_k).]$$

THE FUNDAMENTAL THEOREM OF ALGEBRA

As an application of these notions, we prove the fundamental theorem of algebra: *every nonconstant complex polynomial $P(z)$ must have a zero.*

For the proof it is first necessary to pass from the plane of complex numbers to a compact manifold. Consider the unit sphere $S^2 \subset R^3$ and the stereographic projection

$$h_+ : S^2 - \{(0, 0, 1)\} \to R^2 \times 0 \subset R^3$$

from the "north pole" $(0, 0, 1)$ of S^2. (See Figure 3.) We will identify $R^2 \times 0$ with the plane of complex numbers. The polynomial map P from $R^2 \times 0$ to itself corresponds to a map f from S^2 to itself; where

$$f(x) = h_+^{-1} P h_+(x) \quad \text{for} \quad x \neq (0, 0, 1)$$

$$f(0, 0, 1) = (0, 0, 1).$$

It is well known that this resulting map f is smooth, even in a neighbor-

Fundamental theorem of algebra

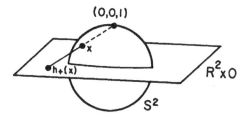

Figure 3. Stereographic projection

hood of the north pole. To see this we introduce the stereographic projection h_- from the south pole $(0, 0, -1)$ and set

$$Q(z) = h_- f h_-^{-1}(z).$$

Note, by elementary geometry, that

$$h_+ h_-^{-1}(z) = z/|z|^2 = 1/\bar{z}.$$

Now if $P(z) = a_0 z^n + a_1 z^{n-1} + \cdots + a_n$, with $a_0 \neq 0$, then a short computation shows that

$$Q(z) = z^n/(\bar{a}_0 + \bar{a}_1 z + \cdots + \bar{a}_n z^n).$$

Thus Q is smooth in a neighborhood of 0, and it follows that $f = h_-^{-1} Q h_-$ is smooth in a neighborhood of $(0, 0, 1)$.

Next observe that f has only a finite number of critical points; for P fails to be a local diffeomorphism only at the zeros of the derivative polynomial $P'(z) = \sum a_{n-j} j z^{j-1}$, and there are only finitely many zeros since P' is not identically zero. The set of regular values of f, being a sphere with finitely many points removed, is therefore connected. Hence the locally constant function $\#f^{-1}(y)$ must actually be constant on this set. Since $\#f^{-1}(y)$ can't be zero everywhere, we conclude that it is zero nowhere. Thus f is an onto mapping, and the polynomial P must have a zero.

§2. THE THEOREM OF SARD AND BROWN

IN GENERAL, it is too much to hope that the set of critical values of a smooth map be finite. But this set will be "small," in the sense indicated by the next theorem, which was proved by A. Sard in 1942 following earlier work by A. P. Morse. (References [30], [24].)

Theorem. *Let* $f : U \to R^n$ *be a smooth map, defined on an open set* $U \subset R^m$, *and let*
$$C = \{x \, \epsilon \, U \mid \text{rank } df_x < n\}.$$
Then the image $f(C) \subset R^n$ *has Lebesgue measure zero.*[*]

Since a set of measure zero cannot contain any nonvacuous open set, it follows that the complement $R^n - f(C)$ must be everywhere dense[†] in R^n.

The proof will be given in §3. It is essential for the proof that f should have many derivatives. (Compare Whitney [38].)

We will be mainly interested in the case $m \geq n$. If $m < n$, then clearly $C = U$; hence the theorem says simply that $f(U)$ has measure zero.

More generally consider a smooth map $f : M \to N$, from a manifold of dimension m to a manifold of dimension n. Let C be the set of all $x \, \epsilon \, M$ such that
$$df_x : TM_x \to TN_{f(x)}$$

---

[*] In other words, given any $\epsilon > 0$, it is possible to cover $f(C)$ by a sequence of cubes in R^n having total n-dimensional volume less than ϵ.

[†] Proved by Arthur B. Brown in 1935. This result was rediscovered by Dubovickiĭ in 1953 and by Thom in 1954. (References [5], [8], [36].)

Regular values

has rank less than n (i.e. is not onto). Then C will be called the set of *critical points*, $f(C)$ the set of *critical values*, and the complement $N - f(C)$ the set of *regular values* of f. (This agrees with our previous definitions in the case $m = n$.) Since M can be covered by a countable collection of neighborhoods each diffeomorphic to an open subset of R^m, we have:

Corollary (A. B. Brown). *The set of regular values of a smooth map $f : M \to N$ is everywhere dense in N.*

In order to exploit this corollary we will need the following:

Lemma 1. *If $f : M \to N$ is a smooth map between manifolds of dimension $m \geq n$, and if $y \in N$ is a regular value, then the set $f^{-1}(y) \subset M$ is a smooth manifold of dimension $m - n$.*

PROOF. Let $x \in f^{-1}(y)$. Since y is a regular value, the derivative df_x must map TM_x onto TN_y. The null space $\mathfrak{N} \subset TM_x$ of df_x will therefore be an $(m - n)$-dimensional vector space.

If $M \subset R^k$, choose a linear map $L : R^k \to R^{m-n}$ that is nonsingular on this subspace $\mathfrak{N} \subset TM_x \subset R^k$. Now define

$$F : M \to N \times R^{m-n}$$

by $F(\xi) = (f(\xi), L(\xi))$. The derivative dF_x is clearly given by the formula

$$dF_x(v) = (df_x(v), L(v)).$$

Thus dF_x is nonsingular. Hence F maps some neighborhood U of x diffeomorphically onto a neighborhood V of $(y, L(x))$. Note that $f^{-1}(y)$ corresponds, under F, to the hyperplane $y \times R^{m-n}$. In fact F maps $f^{-1}(y) \cap U$ diffeomorphically onto $(y \times R^{m-n}) \cap V$. This proves that $f^{-1}(y)$ is a smooth manifold of dimension $m - n$.

As an example we can give an easy proof that the unit sphere S^{m-1} is a smooth manifold. Consider the function $f : R^m \to R$ defined by

$$f(x) = x_1^2 + x_2^2 + \cdots + x_m^2.$$

Any $y \neq 0$ is a regular value, and the smooth manifold $f^{-1}(1)$ is the unit sphere.

If M' is a manifold which is contained in M, it has already been noted that TM'_x is a subspace of TM_x for $x \in M'$. The orthogonal complement of TM'_x in TM_x is then a vector space of dimension $m - m'$ called *the space of normal vectors to M' in M at x.*

In particular let $M' = f^{-1}(y)$ for a regular value y of $f : M \to N$.

Lemma 2. *The null space of $df_x : TM_x \to TN_y$ is precisely equal to the tangent space $TM'_x \subset TM_x$ of the submanifold $M' = f^{-1}(y)$. Hence df_x maps the orthogonal complement of TM'_x isomorphically onto TN_y.*

PROOF. From the diagram

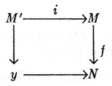

we see that df_x maps the subspace $TM'_x \subset TM_x$ to zero. Counting dimensions we see that df_x maps the space of normal vectors to M' isomorphically onto TN_y.

MANIFOLDS WITH BOUNDARY

The lemmas above can be sharpened so as to apply to a map defined on a smooth "manifold with boundary." Consider first the closed half-space

$$H^m = \{(x_1, \cdots, x_m) \in R^m \mid x_m \geq 0\}.$$

The *boundary* ∂H^m is defined to be the hyperplane $R^{m-1} \times 0 \subset R^m$.

DEFINITION. A subset $X \subset R^k$ is called a *smooth m-manifold with boundary* if each $x \in X$ has a neighborhood $U \cap X$ diffeomorphic to an open subset $V \cap H^m$ of H^m. The *boundary* ∂X is the set of all points in X which correspond to points of ∂H^m under such a diffeomorphism.

It is not hard to show that ∂X is a well-defined smooth manifold of dimension $m - 1$. The *interior* $X - \partial X$ is a smooth manifold of dimension m.

The tangent space TX_x is defined just as in §1, so that TX_x is a full m-dimensional vector space, even if x is a boundary point.

Here is one method for generating examples. Let M be a manifold without boundary and let $g : M \to R$ have 0 as regular value.

Lemma 3. *The set of x in M with $g(x) \geq 0$ is a smooth manifold, with boundary equal to $g^{-1}(0)$.*

The proof is just like the proof of Lemma 1.

EXAMPLE. The *unit disk* D^m, consisting of all $x \in R^m$ with

$$1 - \sum x_i^2 \geq 0,$$

is a smooth manifold, with boundary equal to S^{m-1}.

Now consider a smooth map $f : X \to N$ from an m-manifold with boundary to an n-manifold, where $m > n$.

Lemma 4. *If $y \in N$ is a regular value, both for f and for the restriction $f \mid \partial X$, then $f^{-1}(y) \subset X$ is a smooth $(m - n)$-manifold with boundary. Furthermore the boundary $\partial(f^{-1}(y))$ is precisely equal to the intersection of $f^{-1}(y)$ with ∂X.*

PROOF. Since we have to prove a local property, it suffices to consider the special case of a map $f : H^m \to R^n$, with regular value $y \in R^n$. Let $\bar{x} \in f^{-1}(y)$. If $\bar{x}$ is an interior point, then as before $f^{-1}(y)$ is a smooth manifold in the neighborhood of $\bar{x}$.

Suppose that $\bar{x}$ is a boundary point. Choose a smooth map $g : U \to R^n$ that is defined throughout a neighborhood of $\bar{x}$ in R^m and coincides with f on $U \cap H^m$. Replacing U by a smaller neighborhood if necessary, we may assume that g has no critical points. Hence $g^{-1}(y)$ is a smooth manifold of dimension $m - n$.

Let $\pi : g^{-1}(y) \to R$ denote the coordinate projection,

$$\pi(x_1, \cdots, x_m) = x_m.$$

We claim that π has 0 as a regular value. For the tangent space of $g^{-1}(y)$ at a point $x \in \pi^{-1}(0)$ is equal to the null space of

$$dg_x = df_x : R^m \to R^n;$$

but the hypothesis that $f \mid \partial H^m$ is regular at x guarantees that this null space cannot be completely contained in $R^{m-1} \times 0$.

Therefore the set $g^{-1}(y) \cap H^m = f^{-1}(y) \cap U$, consisting of all $x \in g^{-1}(y)$ with $\pi(x) \geq 0$, is a smooth manifold, by Lemma 3; with boundary equal to $\pi^{-1}(0)$. This completes the proof.

THE BROUWER FIXED POINT THEOREM

We now apply this result to prove the key lemma leading to the classical Brouwer fixed point theorem. Let X be a compact manifold with boundary.

§2. Sard-Brown theorem

Lemma 5. *There is no smooth map $f : X \to \partial X$ that leaves ∂X pointwise fixed.*

PROOF (following M. Hirsch). Suppose there were such a map f. Let $y \, \epsilon \, \partial X$ be a regular value for f. Since y is certainly a regular value for the identity map $f \mid \partial X$ also, it follows that $f^{-1}(y)$ is a smooth 1-manifold, with boundary consisting of the single point

$$f^{-1}(y) \cap \partial X = \{y\}.$$

But $f^{-1}(y)$ is also compact, and the only compact 1-manifolds are finite disjoint unions of circles and segments,* so that $\partial f^{-1}(y)$ must consist of an even number of points. This contradiction establishes the lemma.

In particular the unit disk

$$D^n = \{x \, \epsilon \, R^n \mid x_1^2 + \cdots + x_n^2 \leq 1\}$$

is a compact manifold bounded by the unit sphere S^{n-1}. Hence as a special case we have proved that *the identity map of S^{n-1} cannot be extended to a smooth map $D^n \to S^{n-1}$*.

Lemma 6. *Any smooth map $g : D^n \to D^n$ has a fixed point* (i.e. a point $x \, \epsilon \, D^n$ with $g(x) = x$).

PROOF. Suppose g has no fixed point. For $x \, \epsilon \, D^n$, let $f(x) \, \epsilon \, S^{n-1}$ be the point nearer x than $g(x)$ on the line through x and $g(x)$. (See Figure 4.) Then $f : D^n \to S^{n-1}$ is a smooth map with $f(x) = x$ for $x \, \epsilon \, S^{n-1}$, which is impossible by Lemma 5. (To see that f is smooth we make the following explicit computation: $f(x) = x + tu$, where

$$u = \frac{x - g(x)}{\|x - g(x)\|}, \quad t = -x \cdot u + \sqrt{1 - x \cdot x + (x \cdot u)^2},$$

the expression under the square root sign being strictly positive. Here and subsequently $\|x\|$ denotes the euclidean length $\sqrt{x_1^2 + \cdots + x_n^2}$.)

Brouwer Fixed Point Theorem. *Any continuous function $G : D^n \to D^n$ has a fixed point.*

PROOF. We reduce this theorem to the lemma by approximating G by a smooth mapping. Given $\epsilon > 0$, according to the Weierstrass approximation theorem,† there is a polynomial function $P_1 : R^n \to R^n$ with $\|P_1(x) - G(x)\| < \epsilon$ for $x \, \epsilon \, D^n$. However, P_1 may send points

---

* A proof is given in the Appendix.
† See for example Dieudonné [7, p. 133].

Brouwer fixed point theorem

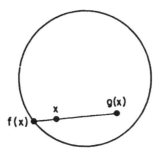

Figure 4

of D^n into points outside of D^n. To correct this we set

$$P(x) = P_1(x)/(1 + \epsilon).$$

Then clearly P maps D^n into D^n and $||P(x) - G(x)|| < 2\epsilon$ for $x \, \epsilon \, D^n$. Suppose that $G(x) \neq x$ for all $x \, \epsilon \, D^n$. Then the continuous function $||G(x) - x||$ must take on a minimum $\mu > 0$ on D^n. Choosing $P : D^n \to D^n$ as above, with $||P(x) - G(x)|| < \mu$ for all x, we clearly have $P(x) \neq x$. Thus P is a smooth map from D^n to itself without a fixed point. This contradicts Lemma 6, and completes the proof.

The procedure employed here can frequently be applied in more general situations: to prove a proposition about continuous mappings, we first establish the result for smooth mappings and then try to use an approximation theorem to pass to the continuous case. (Compare §8, Problem 4.)

§3. PROOF OF SARD'S THEOREM*

FIRST let us recall the statement:

Theorem of Sard. *Let $f : U \to R^p$ be a smooth map, with U open in R^n, and let C be the set of critical points; that is the set of all $x \in U$ with*

$$\operatorname{rank} df_x < p.$$

Then $f(C) \subset R^p$ has measure zero.

REMARK. The cases where $n \leq p$ are comparatively easy. (Compare de Rham [29, p. 10].) We will, however, give a unified proof which makes these cases look just as bad as the others.

The proof will be by induction on n. Note that the statement makes sense for $n \geq 0$, $p \geq 1$. (By definition R^0 consists of a single point.) To start the induction, the theorem is certainly true for $n = 0$.

Let $C_1 \subset C$ denote the set of all $x \in U$ such that the first derivative df_x is zero. More generally let C_i denote the set of x such that all partial derivatives of f of order $\leq i$ vanish at x. Thus we have a descending sequence of closed sets

$$C \supset C_1 \supset C_2 \supset C_3 \supset \cdots .$$

The proof will be divided into three steps as follows:

STEP 1. The image $f(C - C_1)$ has measure zero.
STEP 2. The image $f(C_i - C_{i+1})$ has measure zero, for $i \geq 1$.
STEP 3. The image $f(C_k)$ has measure zero for k sufficiently large.

(REMARK. If f happens to be real analytic, then the intersection of

---

* Our proof is based on that given by Pontryagin [28]. The details are somewhat easier since we assume that f is infinitely differentiable.

Step 1

the C_i is vacuous unless f is constant on an entire component of U. Hence in this case it is sufficient to carry out Steps 1 and 2.)

PROOF OF STEP 1. This first step is perhaps the hardest. We may assume that $p \geq 2$, since $C = C_1$ when $p = 1$. We will need the well known theorem of Fubini* which asserts that *a measurable set*

$$A \subset R^p = R^1 \times R^{p-1}$$

must have measure zero if it intersects each hyperplane (constant) $\times R^{p-1}$ *in a set of $(p-1)$-dimensional measure zero.*

For each $\bar{x} \in C - C_1$ we will find an open neighborhood $V \subset R^n$ so that $f(V \cap C)$ has measure zero. Since $C - C_1$ is covered by countably many of these neighborhoods, this will prove that $f(C - C_1)$ has measure zero.

Since $\bar{x} \notin C_1$, there is some partial derivative, say $\partial f_1/\partial x_1$, which is not zero at $\bar{x}$. Consider the map $h : U \to R^n$ defined by

$$h(x) = (f_1(x), x_2, \cdots, x_n).$$

Since $dh_{\bar{x}}$ is nonsingular, h maps some neighborhood V of $\bar{x}$ diffeomorphically onto an open set V'. The composition $g = f \circ h^{-1}$ will then map V' into R^p. Note that the set C' of critical points of g is precisely $h(V \cap C)$; hence the set $g(C')$ of critical values of g is equal to $f(V \cap C)$.

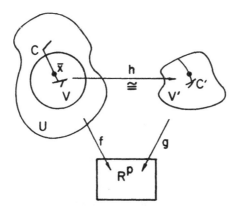

Figure 5. *Construction of the map g*

---

* For an easy proof (as well as an alternative proof of Sard's theorem) see Sternberg [35, pp. 51-52]. Sternberg assumes that A is compact, but the general case follows easily from this special case.

§3. *Proof of Sard's theorem*

For each $(t, x_2, \cdots, x_n) \; \varepsilon \; V'$ note that $g(t, x_2, \cdots, x_n)$ belongs to the hyperplane $t \times R^{p-1} \subset R^p$: thus g carries hyperplanes into hyperplanes. Let

$$g^t : (t \times R^{n-1}) \cap V' \to t \times R^{p-1}$$

denote the restriction of g. Note that a point of $t \times R^{n-1}$ is critical for g^t if and only if it is critical for g; for the matrix of first derivatives of g has the form

$$(\partial g_i/\partial x_j) = \begin{bmatrix} 1 & 0 \\ * & (\partial g_i^t/\partial x_j) \end{bmatrix}.$$

According to the induction hypothesis, the set of critical values of g^t has measure zero in $t \times R^{p-1}$. Therefore the set of critical values of g intersects each hyperplane $t \times R^{p-1}$ in a set of measure zero. This set $g(C')$ is measurable, since it can be expressed as a countable union of compact subsets. Hence, by Fubini's theorem, the set

$$g(C') = f(V \cap C)$$

has measure zero, and Step 1 is complete.

PROOF of STEP 2. For each $\bar{x} \; \varepsilon \; C_k - C_{k+1}$ there is some $(k + 1) - st$ derivative $\partial^{k+1} f_r/\partial x_{s_1} \cdots \partial x_{s_{k+1}}$ which is not zero. Thus the function

$$w(x) = \partial^k f_r/\partial x_{s_1} \cdots \partial x_{s_{k+1}}$$

vanishes at $\bar{x}$ but $\partial w/\partial x_{s_1}$ does not. Suppose for definiteness that $s_1 = 1$. Then the map $h : U \to R^n$ defined by

$$h(x) = (w(x), x_2, \cdots, x_n)$$

carries some neighborhood V of $\bar{x}$ diffeomorphically onto an open set V'. Note that h carries $C_k \cap V$ into the hyperplane $0 \times R^{n-1}$. Again we consider

$$g = f \circ h^{-1} : V' \to R^p.$$

Let

$$\bar{g} : (0 \times R^{n-1}) \cap V' \to R^p$$

denote the restriction of g. By induction, the set of critical values of $\bar{g}$ has measure zero in R^p. But each point in $h(C_k \cap V)$ is certainly a critical point of $\bar{g}$ (since all derivatives of order $\leq k$ vanish). Therefore

$$\bar{g}h(C_k \cap V) = f(C_k \cap V) \text{ has measure zero.}$$

Since $C_k - C_{k+1}$ is covered by countably many such sets V, it follows that $f(C_k - C_{k+1})$ has measure zero.

Step 3

PROOF OF STEP 3. Let $I^n \subset U$ be a cube with edge δ. If k is sufficiently large ($k > n/p - 1$ to be precise) we will prove that $f(C_k \cap I^n)$ has measure zero. Since C_k can be covered by countably many such cubes, this will prove that $f(C_k)$ has measure zero.

From Taylor's theorem, the compactness of I^n, and the definition of C_k, we see that

$$f(x + h) = f(x) + R(x, h)$$

where

1) $$||R(x, h)|| \leq c \, ||h||^{k+1}$$

for $x \, \varepsilon \, C_k \cap I^n$, $x + h \, \varepsilon \, I^n$. Here c is a constant which depends only on f and I^n. Now subdivide I^n into r^n cubes of edge δ/r. Let I_1 be a cube of the subdivision which contains a point x of C_k. Then any point of I_1 can be written as $x + h$, with

2) $$||h|| \leq \sqrt{n}(\delta/r).$$

From 1) it follows that $f(I_1)$ lies in a cube of edge a/r^{k+1} centered about $f(x)$, where $a = 2c \, (\sqrt{n} \, \delta)^{k+1}$ is constant. Hence $f(C_k \cap I^n)$ is contained in a union of at most r^n cubes having total volume

$$V \leq r^n(a/r^{k+1})^p = a^p r^{n-(k+1)p}.$$

If $k + 1 > n/p$, then evidently V tends to 0 as $r \to \infty$; so $f(C_k \cap I^n)$ must have measure zero. This completes the proof of Sard's theorem.

§4. THE DEGREE MODULO 2 OF A MAPPING

CONSIDER a smooth map $f : S^m \to S^m$. If y is a regular value, recall that $\#f^{-1}(y)$ denotes the number of solutions x to the equation $f(x) = y$. We will prove that *the residue class modulo 2 of $\#f^{-1}(y)$ does not depend on the choice of the regular value* y. This residue class is called the mod 2 degree of f. More generally this same definition works for any smooth map

$$f : M \to N$$

where M is compact without boundary, N is connected, and both manifolds have the same dimension. (We may as well assume also that N is compact without boundary, since otherwise the mod 2 degree would necessarily be zero.) For the proof we introduce two new concepts.

SMOOTH HOMOTOPY AND SMOOTH ISOTOPY

Given $X \subset R^k$, let $X \times [0, 1]$ denote the subset* of R^{k+1} consisting of all (x, t) with $x \in X$ and $0 \leq t \leq 1$. Two mappings

$$f, g : X \to Y$$

are called *smoothly homotopic* (abbreviated $f \sim g$) if there exists a

---

* If M is a smooth manifold without boundary, then $M \times [0, 1]$ is a smooth manifold bounded by two "copies" of M. Boundary points of M will give rise to "corner" points of $M \times [0, 1]$.

Homotopy and isotopy

smooth map $F : X \times [0, 1] \to Y$ with
$$F(x, 0) = f(x), \quad F(x, 1) = g(x)$$
for all $x \in X$. This map F is called a *smooth homotopy* between f and g.

Note that the relation of smooth homotopy is an equivalence relation. To see that it is transitive we use the existence of a smooth function $\varphi : [0, 1] \to [0, 1]$ with

$$\varphi(t) = 0 \quad \text{for} \quad 0 \le t \le \tfrac{1}{3}$$
$$\varphi(t) = 1 \quad \text{for} \quad \tfrac{2}{3} \le t \le 1.$$

(For example, let $\varphi(t) = \lambda(t - \tfrac{1}{3})/(\lambda(t - \tfrac{1}{3}) + \lambda(\tfrac{2}{3} - t))$, where $\lambda(\tau) = 0$ for $\tau \le 0$ and $\lambda(\tau) = \exp(-\tau^{-1})$ for $\tau > 0$.) Given a smooth homotopy F between f and g, the formula $G(x, t) = F(x, \varphi(t))$ defines a smooth homotopy G with

$$G(x, t) = f(x) \quad \text{for} \quad 0 \le t \le \tfrac{1}{3}$$
$$G(x, t) = g(x) \quad \text{for} \quad \tfrac{2}{3} \le t \le 1.$$

Now if $f \sim g$ and $g \sim h$, then, with the aid of this construction, it is easy to prove that $f \sim h$.

If f and g happen to be diffeomorphisms from X to Y, we can also define the concept of a "smooth isotopy" between f and g. This also will be an equivalence relation.

DEFINITION. The diffeomorphism f is *smoothly isotopic* to g if there exists a smooth homotopy $F : X \times [0, 1] \to Y$ from f to g so that, for each $t \in [0, 1]$, the correspondence

$$x \mapsto F(x, t)$$

maps X diffeomorphically onto Y.

It will turn out that the mod 2 degree of a map depends only on its smooth homotopy class:

Homotopy Lemma. *Let $f, g : M \to N$ be smoothly homotopic maps between manifolds of the same dimension, where M is compact and without boundary. If $y \in N$ is a regular value for both f and g, then*

$$\#f^{-1}(y) \equiv \#g^{-1}(y) \pmod{2}.$$

PROOF. Let $F : M \times [0, 1] \to N$ be a smooth homotopy between f and g. First suppose that y is also a regular value for F. Then $F^{-1}(y)$

is a compact 1-manifold, with boundary equal to

$$F^{-1}(y) \cap (M \times 0 \cup M \times 1) = f^{-1}(y) \times 0 \cup g^{-1}(y) \times 1.$$

Thus the total number of boundary points of $F^{-1}(y)$ is equal to

$$\#f^{-1}(y) + \#g^{-1}(y).$$

But we recall from §2 that a compact 1-manifold always has an even number of boundary points. Thus $\#f^{-1}(y) + \#g^{-1}(y)$ is even, and therefore

$$\#f^{-1}(y) \equiv \#g^{-1}(y) \pmod{2}.$$

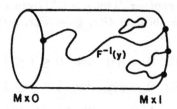

Figure 6. *The number of boundary points on the left is congruent to the number on the right modulo 2*

Now suppose that y is not a regular value of F. Recall (from §1) that $\#f^{-1}(y')$ and $\#g^{-1}(y')$ are locally constant functions of y' (as long as we stay away from critical values). Thus there is a neighborhood $V_1 \subset N$ of y, consisting of regular values of f, so that

$$\#f^{-1}(y') = \#f^{-1}(y)$$

for all $y' \in V_1$; and there is an analogous neighborhood $V_2 \subset N$ so that

$$\#g^{-1}(y') = \#g^{-1}(y)$$

for all $y' \in V_2$. Choose a regular value z of F within $V_1 \cap V_2$. Then

$$\#f^{-1}(y) = \#f^{-1}(z) \equiv \#g^{-1}(z) = \#g^{-1}(y)),$$

which completes the proof.

We will also need the following:

Homogeneity Lemma. *Let y and z be arbitrary interior points of the smooth, connected manifold N. Then there exists a diffeomorphism $h: N \to N$ that is smoothly isotopic to the identity and carries y into z.*

Homotopy and isotopy

(For the special case $N = S^n$ the proof is easy: simply choose h to be the rotation which carries y into z and leaves fixed all vectors orthogonal to the plane through y and z.)

The proof in general proceeds as follows: We will first construct a smooth isotopy from R^n to itself which

1) leaves all points outside of the unit ball fixed, and
2) slides the origin to any desired point of the open unit ball.

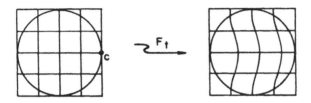

Figure 7. *Deforming the unit ball*

Let $\varphi : R^n \to R$ be a smooth function which satisfies

$$\varphi(x) > 0 \quad \text{for} \quad ||x|| < 1$$
$$\varphi(x) = 0 \quad \text{for} \quad ||x|| \geq 1.$$

(For example let $\varphi(x) = \lambda(1 - ||x||^2)$ where $\lambda(t) = 0$ for $t \leq 0$ and $\lambda(t) = \exp(-t^{-1})$ for $t > 0$.) Given any fixed unit vector $c \in S^{n-1}$, consider the differential equations

$$\frac{dx_i}{dt} = c_i \varphi(x_1, \cdots, x_n); \quad i = 1, \cdots, n.$$

For any $\bar{x} \in R^n$ these equations have a unique solution $x = x(t)$, defined for all* real numbers, which satisfies the initial condition

$$x(0) = \bar{x}.$$

We will use the notation $x(t) = F_t(\bar{x})$ for this solution. Then clearly

1) $F_t(\bar{x})$ is defined for all t and $\bar{x}$ and depends smoothly on t and $\bar{x}$,
2) $F_0(\bar{x}) = \bar{x}$,
3) $F_{s+t}(\bar{x}) = F_s \circ F_t(\bar{x})$.

---

* Compare [22, §2.4].

§4. Degree modulo 2

Therefore each F_t is a diffeomorphism from R^n onto itself. Letting t vary, we see that each F_t is smoothly isotopic to the identity under an isotopy which leaves all points outside of the unit ball fixed. But clearly, with suitable choice of c and t, the diffeomorphism F_t will carry the origin to any desired point in the open unit ball.

Now consider a connected manifold N. Call two points of N "isotopic" if there exists a smooth isotopy carrying one to the other. This is clearly an equivalence relation. If y is an interior point, then it has a neighborhood diffeomorphic to R^n; hence the above argument shows that every point sufficiently close to y is "isotopic" to y. In other words, each "isotopy class" of points in the interior of N is an open set, and the interior of N is partitioned into disjoint open isotopy classes. But the interior of N is connected; hence there can be only one such isotopy class. This completes the proof.

We can now prove the main result of this section. Assume that M is compact and boundaryless, that N is connected, and that $f : M \to N$ is smooth.

Theorem. *If y and z are regular values of f then*

$$\#f^{-1}(y) \equiv \#f^{-1}(z) \quad (\text{modulo } 2).$$

This common residue class, which is called the mod 2 *degree of f, depends only on the smooth homotopy class of f.*

PROOF. Given regular values y and z, let h be a diffeomorphism from N to N which is isotopic to the identity and which carries y to z. Then z is a regular value of the composition $h \circ f$. Since $h \circ f$ is homotopic to f, the Homotopy Lemma asserts that

$$\#(h \circ f)^{-1}(z) \equiv \#f^{-1}(z) \quad (\text{mod } 2).$$

But

$$(h \circ f)^{-1}(z) = f^{-1}h^{-1}(z) = f^{-1}(y),$$

so that

$$\#(h \circ f)^{-1}(z) = \#f^{-1}(y).$$

Therefore

$$\#f^{-1}(y) \equiv \#f^{-1}(z) \quad (\text{mod } 2),$$

as required.

Call this common residue class $\deg_2(f)$. Now suppose that f is smoothly homotopic to g. By Sard's theorem, there exists an element $y \in N$

Homotopy and isotopy 25

which is a regular value for both f and g. The congruence

$$\deg_2 f \equiv \#f^{-1}(y) \equiv \#g^{-1}(y) \equiv \deg_2 g \quad (\text{mod } 2)$$

now shows that $\deg_2 f$ is a smooth homotopy invariant, and completes the proof.

EXAMPLES. A constant map $c : M \to M$ has even mod 2 degree. The identity map I of M has odd degree. *Hence the identity map of a compact boundaryless manifold is not homotopic to a constant.*

In the case $M = S^n$, this result implies the assertion that no smooth map $f : D^{n+1} \to S^n$ leaves the sphere pointwise fixed. (I.e., the sphere is not a smooth "retract" of the disk. Compare §2, Lemma 5.) For such a map f would give rise to a smooth homotopy

$$F : S^n \times [0, 1] \to S^n, \quad F(x, t) = f(tx),$$

between a constant map and the identity.

§5. ORIENTED MANIFOLDS

IN ORDER to define the degree as an integer (rather than an integer modulo 2) we must introduce orientations.

DEFINITIONS. An orientation for a finite dimensional real vector space is an equivalence class of ordered bases as follows: the ordered basis $(b_1, \cdots, b_n)$ determines the *same orientation* as the basis $(b'_1, \cdots, b'_n)$ if $b'_i = \sum a_{ij}b_j$ with $\det(a_{ij}) > 0$. It determines the *opposite orientation* if $\det(a_{ij}) < 0$. Thus each positive dimensional vector space has precisely two orientations. The vector space R^n has a *standard* orientation corresponding to the basis $(1, 0, \cdots, 0), (0, 1, 0, \cdots, 0), \cdots, (0, \cdots, 0, 1)$.

In the case of the zero dimensional vector space it is convenient to define an "orientation" as the symbol $+1$ or -1.

An *oriented* smooth manifold consists of a manifold M together with a choice of orientation for each tangent space TM_x. If $m \geq 1$, these are required to fit together as follows: For each point of M there should exist a neighborhood $U \subset M$ and a diffeomorphism h mapping U onto an open subset of R^m or H^m which is *orientation preserving*, in the sense that for each $x \in U$ the isomorphism dh_x carries the specified orientation for TM_x into the standard orientation for R^m.

If M is connected and orientable, then it has precisely two orientations.

If M has a boundary, we can distinguish three kinds of vectors in the tangent space TM_x at a boundary point:

1) there are the vectors tangent to the boundary, forming an $(m - 1)$-dimensional subspace $T(\partial M)_x \subset TM_x$;

2) there are the "outward" vectors, forming an open half space bounded by $T(\partial M)_x$;

3) there are the "inward" vectors forming a complementary half space.

Each orientation for M determines an orientation for ∂M as follows: For $x \in \partial M$ choose a positively oriented basis $(v_1, v_2, \cdots, v_m)$ for TM_x in such a way that $v_2, \cdots, v_m$ are tangent to the boundary (assuming that $m \geq 2$) and that v_1 is an "outward" vector. Then $(v_2, \cdots, v_m)$ determines the required orientation for ∂M at x.

If the dimension of M is 1, then each boundary point x is assigned the orientation -1 or $+1$ according as a positively oriented vector at x points inward or outward. (See Figure 8.)

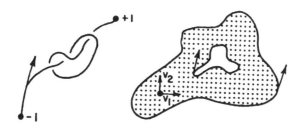

Figure 8. How to orient a boundary

As an example the unit sphere $S^{m-1} \subset R^m$ can be oriented as the boundary of the disk D^m.

THE BROUWER DEGREE

Now let M and N be oriented n-dimensional manifolds without boundary and let

$$f : M \to N$$

be a smooth map. If M is compact and N is connected, then the degree of f is defined as follows:

Let $x \in M$ be a regular point of f, so that $df_x : TM_x \to TN_{f(x)}$ is a linear isomorphism between oriented vector spaces. Define the *sign* of df_x to be $+1$ or -1 according as df_x preserves or reverses orientation. For any regular value $y \in N$ define

$$\deg(f; y) = \sum_{x \in f^{-1}(y)} \operatorname{sign} df_x.$$

As in §1, this integer $\deg(f; y)$ is a locally constant function of y. It is defined on a dense open subset of N.

§5. *Oriented manifolds*

Theorem A. *The integer* $\deg(f; y)$ *does not depend on the choice of regular value* y.

It will be called the *degree* of f (denoted $\deg f$).

Theorem B. *If f is smoothly homotopic to g, then* $\deg f = \deg g$.

The proof will be essentially the same as that in §4. It is only necessary to keep careful control of orientations.

First consider the following situation: Suppose that M is the boundary of a compact oriented manifold X and that M is oriented as the boundary of X.

Lemma 1. *If $f : M \to N$ extends to a smooth map $F : X \to N$, then* $\deg(f; y) = 0$ *for every regular value* y.

PROOF. First suppose that y is a regular value for F, as well as for $f = F \mid M$. The compact 1-manifold $F^{-1}(y)$ is a finite union of arcs and circles, with only the boundary points of the arcs lying on $M = \partial X$. Let $A \subset F^{-1}(y)$ be one of these arcs, with $\partial A = \{a\} \cup \{b\}$. We will show that

$$\text{sign } df_a + \text{sign } df_b = 0,$$

and hence (summing over all such arcs) that $\deg(f ; y) = 0$.

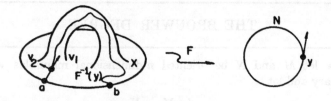

Figure 9. *How to orient* $F^{-1}(y)$

The orientations for X and N determine an orientation for A as follows: Given $x \in A$, let $(v_1, \cdots, v_{n+1})$ be a positively oriented basis for TX_x with v_1 tangent to A. Then v_1 determines the required orientation for TA_x if and only if dF_x carries $(v_2, \cdots, v_{n+1})$ into a positively oriented basis for TN_y.

Let $v_1(x)$ denote the positively oriented unit vector tangent to A at x. Clearly v_1 is a smooth function, and $v_1(x)$ points outward at one boundary point (say b) and inward at the other boundary point a.

It follows immediately that

$$\text{sign } df_a = -1, \quad \text{sign } df_b = +1;$$

with sum zero. Adding up over all such arcs A, we have proved that $\deg(f; y) = 0$.

More generally, suppose that y_0 is a regular value for f, but not for F. The function $\deg(f; y)$ is constant within some neighborhood U of y_0. Hence, as in §4, we can choose a regular value y for F within U and observe that

$$\deg(f; y_0) = \deg(f; y) = 0.$$

This proves Lemma 1.

Now consider a smooth homotopy $\quad F : [0, 1] \times M \to N \quad$ between two mappings $\quad f(x) = F(0, x), \quad g(x) = F(1, x)$.

Lemma 2. *The degree* $\deg(g; y)$ *is equal to* $\deg(f; y)$ *for any common regular value* y.

PROOF. The manifold $[0, 1] \times M^n$ can be oriented as a product, and will then have boundary consisting of $1 \times M^n$ (with the correct orientation) and $0 \times M^n$ (with the wrong orientation). Thus the degree of $F \mid \partial([0, 1] \times M^n)$ at a regular value y is equal to the difference

$$\deg(g; y) - \deg(f; y).$$

According to Lemma 1 this difference must be zero.

The remainder of the proof of Theorems A and B is completely analogous to the argument in §4. If y and z are both regular values for $f : M \to N$, choose a diffeomorphism $h : N \to N$ that carries y to z and is isotopic to the identity. Then h will preserve orientation, and

$$\deg(f; y) = \deg(h \circ f; h(y))$$

by inspection. But f is homotopic to $h \circ f$; hence

$$\deg(h \circ f; z) = \deg(f; z)$$

by Lemma 2. Therefore $\deg(f; y) = \deg(f; z)$, which completes the proof.

EXAMPLES. The complex function $z \to z^k$, $z \neq 0$, maps the unit circle onto itself with degree k. (Here k may be positive, negative, or zero.) The degenerate mapping

$$f : M \to \text{constant} \; \epsilon \; N$$

§5. Oriented manifolds

has degree zero. A diffeomorphism $f : M \to N$ has degree $+1$ or -1 according as f preserves or reverses orientation. *Thus an orientation reversing diffeomorphism of a compact boundaryless manifold is not smoothly homotopic to the identity.*

One example of an orientation reversing diffeomorphism is provided by the reflection $r_i : S^n \to S^n$, where

$$r_i(x_1, \cdots, x_{n+1}) = (x_1, \cdots, -x_i, \cdots, x_{n+1}).$$

The antipodal map of S^n has degree $(-1)^{n+1}$, as we can see by noting that the antipodal map is the composition of $n + 1$ reflections:

$$-x = r_1 \circ r_2 \circ \cdots \circ r_{n+1}(x).$$

Thus if n is even, the antipodal map of S^n is not smoothly homotopic to the identity, a fact not detected by the degree modulo 2.

As an application, following Brouwer, we show that S^n admits a smooth field of nonzero tangent vectors if and only if n is odd. (Compare Figures 10 and 11.)

Figure 10 (above). A nonzero vector field on the 1-sphere

Figure 11 (below). Attempts for $n = 2$

DEFINITION. A smooth *tangent vector field* on $M \subset R^k$ is a smooth map $v : M \to R^k$ such that $v(x) \in TM_x$ for each $x \in M$. In the case of the sphere $S^n \subset R^{n+1}$ this is clearly equivalent to the condition

1) $$v(x) \cdot x = 0 \quad \text{for all} \quad x \in S^n,$$

The Brouwer degree

using the euclidean inner product.

If $v(x)$ is nonzero for all x, then we may as well suppose that

2) $\qquad v(x) \cdot v(x) = 1 \quad \text{for all} \quad x \in S^n.$

For in any case $\bar{v}(x) = v(x)/\|v(x)\|$ would be a vector field which does satisfy this condition. Thus we can think of v as a smooth function from S^n to itself.

Now define a smooth homotopy

$$F : S^n \times [0, \pi] \to S^n$$

by the formula $F(x, \theta) = x \cos \theta + v(x) \sin \theta$. Computation shows that

$$F(x, \theta) \cdot F(x, \theta) = 1$$

and that

$$F(x, 0) = x, \qquad F(x, \pi) = -x.$$

Thus the antipodal map of S^n is homotopic to the identity. But for n even we have seen that this is impossible.

On the other hand, if $n = 2k - 1$, the explicit formula

$$v(x_1, \cdots, x_{2k}) = (x_2, -x_1, x_4, -x_3, \cdots, x_{2k}, -x_{2k-1})$$

defines a nonzero tangent vector field on S^n. This completes the proof.

It follows, incidentally, that the antipodal map of S^n *is* homotopic to the identity for n odd. A famous theorem due to Heinz Hopf asserts that two mappings from a connected n-manifold to the n-sphere are smoothly homotopic *if and only if* they have the same degree. In §7 we will prove a more general result which implies Hopf's theorem.

§6. VECTOR FIELDS AND THE EULER NUMBER

As a further application of the concept of degree, we study vector fields on other manifolds.

Consider first an open set $U \subset R^m$ and a smooth vector field

$$v : U \to R^m$$

with an isolated zero at the point $z \, \varepsilon \, U$. The function

$$\bar{v}(x) = v(x)/||v(x)||$$

maps a small sphere centered at z into the unit sphere.* The degree of this mapping is called the *index* ι of v at the zero z.

Some examples, with indices $-1, 0, 1, 2$, are illustrated in Figure 12. (Intimately associated with v are the curves "tangent" to v which are obtained by solving the differential equations $dx_i/dt = v_i(x_1, \cdots, x_n)$. It is these curves which are actually sketched in Figure 12.)

A zero with arbitrary index can be obtained as follows: In the plane of complex numbers the polynomial z^k defines a smooth vector field with a zero of index k at the origin, and the function $\bar{z}^k$ defines a vector field with a zero of index $-k$.

We must prove that this concept of index is invariant under diffeomorphism of U. To explain what this means, let us consider the more general situation of a map $f : M \to N$, with a vector field on each manifold.

DEFINITION. The vector fields v on M and v' on N *correspond* under f if the derivative df_x carries $v(x)$ into $v'(f(x))$ for each $x \, \varepsilon \, M$.

---

* Each sphere is to be oriented as the boundary of the corresponding disk.

The index

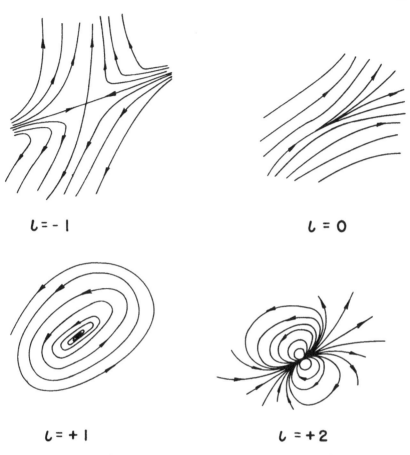

$\iota = -1$ $\iota = 0$

$\iota = +1$ $\iota = +2$

Figure 12. Examples of plane vector fields

If f is a diffeomorphism, then clearly v' is uniquely determined by v. The notation

$$v' = df \circ v \circ f^{-1}$$

will be used.

Lemma 1. *Suppose that the vector field v on U corresponds to*

$$v' = df \circ v \circ f^{-1}$$

on U' under a diffeomorphism $f : U \to U'$. Then the index of v at an isolated zero z is equal to the index of v' at $f(z)$.

Assuming Lemma 1, we can define the concept of index for a vector field w on an arbitrary manifold M as follows: If $g : U \to M$ is a parametrization of a neighborhood of z in M, then the *index* ι of w at z is defined to be equal to the index of the corresponding vector field $dg^{-1} \circ w \circ g$ on U at the zero $g^{-1}(z)$. It clearly will follow from Lemma 1 that ι is well defined.

The proof of Lemma 1 will be based on the proof of a quite different result:

Lemma 2. *Any orientation preserving diffeomorphism f of R^m is smoothly isotopic to the identity.*

(In contrast, for many values of m there exists an orientation preserving diffeomorphism of the sphere S^m which is not smoothly isotopic to the identity. See [20, p. 404].)

PROOF. We may assume that $f(0) = 0$. Since the derivative at 0 can be defined by

$$df_0(x) = \lim_{t \to 0} f(tx)/t,$$

it is natural to define an isotopy

$$F : R^m \times [0, 1] \to R^m$$

by the formula

$$F(x, t) = f(tx)/t \quad \text{for} \quad 0 < t \leq 1,$$
$$F(x, 0) = df_0(x).$$

To prove that F is smooth, even as $t \to 0$, we write f in the form*

$$f(x) = x_1 g_1(x) + \cdots + x_m g_m(x),$$

where $g_1, \cdots, g_m$ are suitable smooth functions, and note that

$$F(x, t) = x_1 g_1(tx) + \cdots + x_m g_m(tx)$$

for all values of t.

Thus f is isotopic to the linear mapping df_0, which is clearly isotopic to the identity. This proves Lemma 2.

PROOF OF LEMMA 1. We may assume that $z = f(z) = 0$ and that U is convex. If f preserves orientation, then, proceeding exactly as above,

---
* See for example [22, p. 5].

we construct a one-parameter family of embeddings

$$f_t : U \to R^m$$

with f_0 = identity, $f_1 = f$, and $f_t(0) = 0$ for all t. Let v_t denote the vector field $df_t \circ v \circ f_t^{-1}$ on $f_t(U)$, which corresponds to v on U. These vector fields are all defined and nonzero on a sufficiently small sphere centered at 0. Hence the index of $v = v_0$ at 0 must be equal to the index of $v' = v_1$ at 0. This proves Lemma 1 for orientation preserving diffeomorphisms.

To consider diffeomorphisms which reverse orientation it is sufficient to consider the special case of a reflection ρ. Then

$$v' = \rho \circ v \circ \rho^{-1},$$

so the associated function $\bar{v}'(x) = v'(x)/||v'(x)||$ on the ϵ-sphere satisfies

$$\bar{v}' = \rho \circ \bar{v} \circ \rho^{-1}.$$

Evidently the degree of $\bar{v}'$ equals the degree of $\bar{v}$, which completes the proof of Lemma 1.

We will study the following classical result: Let M be a compact manifold and w a smooth vector field on M with isolated zeros. *If M has a boundary, then w is required to point outward at all boundary points.*

Poincaré-Hopf Theorem. *The sum $\sum \iota$ of the indices at the zeros of such a vector field is equal to the Euler number*[*]

$$\chi(M) = \sum_{i=0}^{m} (-1)^i \text{ rank } H_i(M).$$

In particular this index sum is a topological invariant of M: it does not depend on the particular choice of vector field.

(A 2-dimensional version of this theorem was proved by Poincaré in 1885. The full theorem was proved by Hopf [14] in 1926 after earlier partial results by Brouwer and Hadamard.)

We will prove part of this theorem, and sketch a proof of the rest. First consider the special case of a compact domain in R^m.

Let $X \subset R^m$ be a compact m-manifold with boundary. The *Gauss mapping*

$$g : \partial X \to S^{m-1}$$

assigns to each $x \in \partial X$ the outward unit normal vector at x.

---

[*] Here $H_i(M)$ denotes the i-th homology group of M. This will be our first and last reference to homology theory.

§6. Vector fields

Lemma 3 (Hopf). *If $v : X \to R^m$ is a smooth vector field with isolated zeros, and if v points out of X along the boundary, then the index sum $\sum \iota$ is equal to the degree of the Gauss mapping from ∂X to S^{m-1}. In particular, $\sum \iota$ does not depend on the choice of v.*

For example, if a vector field on the disk D^m points outward along the boundary, then $\sum \iota = +1$. (Compare Figure 13.)

Figure 13. An example with index sum $+1$

PROOF. Removing an ϵ-ball around each zero, we obtain a new manifold with boundary. The function $\bar{v}(x) = v(x)/\|v(x)\|$ maps this manifold into S^{m-1}. Hence the sum of the degrees of $\bar{v}$ restricted to the various boundary components is zero. But $\bar{v} \mid \partial X$ is homotopic to g, and the degrees on the other boundary components add up to $-\sum \iota$. (The minus sign occurs since each small sphere gets the wrong orientation.) Therefore

$$\deg(g) - \sum \iota = 0$$

as required.

REMARK. The degree of g is also known as the "curvatura integra" of ∂X, since it can be expressed as a constant times the integral over ∂X of the Gaussian curvature. This integer is of course equal to the Euler number of X. For m odd it is equal to half the Euler number of ∂X.

Before extending this result to other manifolds, some more preliminaries are needed.

It is natural to try to compute the index of a vector field v at a zero z

The index sum

in terms of the derivatives of v at z. Consider first a vector field v on an open set $U \subset R^m$ and think of v as a mapping $U \to R^m$, so that $dv_z : R^m \to R^m$ is defined.

DEFINITION. The vector field v is *nondegenerate* at z if the linear transformation dv_z is nonsingular.

It follows that z is an isolated zero.

Lemma 4. *The index of v at a nondegenerate zero z is either $+1$ or -1 according as the determinant of dv_z is positive or negative.*

PROOF. Think of v as a diffeomorphism from some convex neighborhood U_0 of z into R^m. We may assume that $z = 0$. If v preserves orientation, we have seen that $v|U_0$ can be deformed smoothly into the identity without introducing any new zeros. (See Lemmas 1, 2.) Hence the index is certainly equal to $+1$.

If v reverses orientation, then similarly v can be deformed into a reflection; hence $\iota = -1$.

More generally consider a zero z of a vector field w on a manifold $M \subset R^k$. Think of w as a map from M to R^k so that the derivative $dw_z : TM_z \to R^k$ is defined.

Lemma 5. *The derivative dw_z actually carries TM_z into the subspace $TM_z \subset R^k$, and hence can be considered as a linear transformation from TM_z to itself. If this linear transformation has determinant $D \neq 0$ then z is an isolated zero of w with index equal to $+1$ or -1 according as D is positive or negative.*

PROOF. Let $h : U \to M$ be a parametrization of some neighborhood of z. Let e^i denote the i-th basis vector of R^m and let

$$t^i = dh_u(e^i) = \partial h/\partial u_i$$

so that the vectors $t^1, \cdots, t^m$ form a basis for the tangent space $TM_{h(u)}$. We must compute the image of $t^i = t^i(u)$ under the linear transformation $dw_{h(u)}$. First note that

1) $\qquad dw_{h(u)}(t^i) = d(w \circ h)_u(e^i) = \partial w(h(u))/\partial u_i .$

Let $v = \sum v_i e^i$ be the vector field on U which corresponds to the vector field w on M. By definition $v = dh^{-1} \circ w \circ h$, so that

$$w(h(u)) = dh_u(v) = \sum v_i t^i .$$

Therefore

2) $\qquad \partial w(h(u))/\partial u_i = \sum_i (\partial v_i/\partial u_i) t^i + \sum_i v_i(\partial t^i/\partial u_i).$

Combining 1) and 2), and then evaluating at the zero $h^{-1}(z)$ of v, we obtain the formula

3) $$dw_z(t^i) = \sum_i (\partial v_i/\partial u_i)t^i.$$

Thus dw_z maps TM_z into itself, and the determinant D of this linear transformation $TM_z \to TM_z$ is equal to the determinant of the matrix $(\partial v_i/\partial u_i)$. Together with Lemma 4 this completes the proof.

Now consider a compact, boundaryless manifold $M \subset R^k$. Let N_ϵ denote the closed ϵ-neighborhood of M(i.e., the set of all $x \in R^k$ with $\|x - y\| \leq \epsilon$ for some $y \in M$). For ϵ sufficiently small one can show that N_ϵ is a smooth manifold with boundary. (See §8, Problem 11.)

Theorem 1. *For any vector field v on M with only nondegenerate zeros, the index sum $\sum \iota$ is equal to the degree of the Gauss mapping**

$$g : \partial N_\epsilon \to S^{k-1}.$$

In particular this sum does not depend on the choice of vector field.

PROOF. For $x \in N_\epsilon$ let $r(x) \in M$ denote the closest point of M. (Compare §8, Problem 12.) Note that the vector $x - r(x)$ is perpendicular to the tangent space of M at $r(x)$, for otherwise $r(x)$ would not be the closest point of M. If ϵ is sufficiently small, then the function $r(x)$ is smooth and well defined.

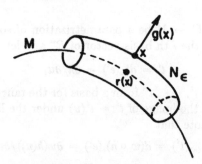

Figure 14. The ϵ-neighborhood of M

---

* A different interpretation of this degree has been given by Allendoerfer and Fenchel: the degree of g can be expressed as the integral over M of a suitable curvature scalar, thus yielding an m-dimensional version of the classical Gauss-Bonnet theorem. (References [1], [9]. See also Chern [6].)

We will also consider the squared distance function

$$\varphi(x) = \|x - r(x)\|^2.$$

An easy computation shows that the gradient of φ is given by

$$\operatorname{grad} \varphi = 2(x - r(x)).$$

Hence, for each point x of the level surface $\partial N_\epsilon = \varphi^{-1}(\epsilon^2)$, the outward unit normal vector is given by

$$g(x) = \operatorname{grad} \varphi / \|\operatorname{grad} \varphi\| = (x - r(x))/\epsilon.$$

Extend v to a vector field w on the neighborhood N_ϵ by setting

$$w(x) = (x - r(x)) + v(r(x)).$$

Then w points outward along the boundary, since the inner product $w(x) \cdot g(x)$ is equal to $\epsilon > 0$. Note that w can vanish only at the zeros of v in M; this is clear since the two summands $(x - r(x))$ and $v(r(x))$ are mutually orthogonal. Computing the derivative of w at a zero $z \in M$, we see that

$$dw_z(h) = dv_z(h) \quad \text{for all} \quad h \in TM_z$$
$$dw_z(h) = h \quad \text{for} \quad h \in TM_z^\perp.$$

Thus the determinant of dw_z is equal to the determinant of dv_z. Hence the index of w at the zero z is equal to the index ι of v at z.

Now according to Lemma 3 the index sum $\sum \iota$ is equal to the degree of g. This proves Theorem 1.

EXAMPLES. On the sphere S^m there exists a vector field v which points "north" at every point.* At the south pole the vectors radiate outward; hence the index is $+1$. At the north pole the vectors converge inward; hence the index is $(-1)^m$. Thus the invariant $\sum \iota$ is equal to 0 or 2 according as m is odd or even. This gives a new proof that every vector field on an even sphere has a zero.

For any odd-dimensional, boundaryless manifold the invariant $\sum \iota$ is zero. For if the vector field v is replaced by $-v$, then each index is multiplied by $(-1)^m$, and the equality

$$\sum \iota = (-1)^m \sum \iota,$$

for m odd, implies that $\sum \iota = 0$.

---

* For example, v can be defined by the formula $v(x) = p - (p \cdot x)x$, where p is the north pole. (See Figure 11.)

§6. *Vector fields*

REMARK. If $\sum \iota = 0$ on a connected manifold M, then a theorem of Hopf asserts that there exists a vector field on M with no zeros at all.

In order to obtain the full strength of the Poincaré-Hopf theorem, three further steps are needed.

STEP 1. *Identification of the invariant $\sum \iota$ with the Euler number $\chi(M)$.* It is sufficient to construct just one example of a nondegenerate vector field on M with $\sum \iota$ equal to $\chi(M)$. The most pleasant way of doing this is the following: According to Marston Morse, it is always possible to find a real valued function on M whose "gradient" is a nondegenerate vector field. Furthermore, Morse showed that the sum of indices associated with such a gradient field is equal to the Euler number of M. For details of this argument the reader is referred to Milnor [22, pp. 29, 36].

STEP 2. *Proving the theorem for a vector field with degenerate zeros.* Consider first a vector field v on an open set U with an isolated zero at z. If

$$\lambda : U \to [0, 1]$$

takes the value 1 on a small neighborhood N_1 of z and the value 0 outside a slightly larger neighborhood N, and if y is a sufficiently small regular value of v, then the vector field

$$v'(x) = v(x) - \lambda(x)y$$

is nondegenerate* within N. The sum of the indices at the zeros within N can be evaluated as the degree of the map

$$\bar{v} : \partial N \to S^{m-1},$$

and hence does not change during this alteration.

More generally consider vector fields on a compact manifold M. Applying this argument locally we see that *any vector field with isolated zeros can be replaced by a nondegenerate vector field without altering the integer $\sum \iota$.*

STEP 3. *Manifolds with boundary.* If $M \subset R^k$ has a boundary, then any vector field v which points outward along ∂M can again be extended over the neighborhood N_ϵ so as to point outward along ∂N_ϵ. However, there is some difficulty with smoothness around the boundary of M. Thus N_ϵ is not a smooth (i.e. differentiable of class C^∞) manifold,

---

* Clearly v' is nondegenerate within N_1. But if y is sufficiently small, then v' will have no zeros at all within $N - N_1$.

The index sum

but only a C^1-manifold. The extension w, if defined as before by $w(x) = v(r(x)) + x - r(x)$, will only be a continuous vector field near ∂M. The argument can nonetheless be carried out either by showing that our strong differentiability assumptions are not really necessary or by other methods.

§7. FRAMED COBORDISM
THE PONTRYAGIN CONSTRUCTION

THE degree of a mapping $M \to M'$ is defined only when the manifolds M and M' are oriented and have the same dimension. We will study a generalization, due to Pontryagin, which is defined for a smooth map

$$f : M \to S^p$$

from an arbitrary compact, boundaryless manifold to a sphere. First some definitions.

Let N and N' be compact n-dimensional submanifolds of M with $\partial N = \partial N' = \partial M = \varnothing$. The difference of dimensions $m - n$ is called the *codimension* of the submanifolds.

DEFINITION. N is *cobordant* to N' *within* M if the subset

$$N \times [0, \epsilon) \cup N' \times (1 - \epsilon, 1]$$

of $M \times [0, 1]$ can be extended to a compact manifold

$$X \subset M \times [0, 1]$$

so that

$$\partial X = N \times 0 \cup N' \times 1,$$

and so that X does not intersect $M \times 0 \cup M \times 1$ except at the points of ∂X.

Clearly cobordism is an equivalence relation. (See Figure 15.)

DEFINITION. A *framing* of the submanifold $N \subset M$ is a smooth function $\mathfrak{v}$ which assigns to each $x \in N$ a basis

$$\mathfrak{v}(x) = (v^1(x), \cdots, v^{m-n}(x))$$

for the space $TN_x^\perp \subset TM_x$ of normal vectors to N in M at x. (See

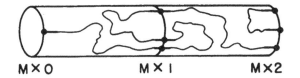

M × 0   M × 1   M × 2

Figure 15. Pasting together two cobordisms within M

Figure 16.) The pair $(N, \mathfrak{v})$ is called a *framed submanifold* of M. Two framed submanifolds $(N, \mathfrak{v})$ and $(N', \mathfrak{w})$ are *framed cobordant* if there exists a cobordism $X \subset M \times [0, 1]$ between N and N' and a framing $\mathfrak{u}$ of X, so that

$$u^i(x, t) = (v^i(x), 0) \quad \text{for} \quad (x, t) \, \varepsilon \, N \times [0, \epsilon)$$
$$u^i(x, t) = (w^i(x), 0) \quad \text{for} \quad (x, t) \, \varepsilon \, N' \times (1 - \epsilon, 1].$$

Again this is an equivalence relation.

Now consider a smooth map $f : M \to S^p$ and a regular value $y \, \varepsilon \, S^p$. The map f induces a framing of the manifold $f^{-1}(y)$ as follows: Choose a positively oriented basis $\mathfrak{v} = (v^1, \cdots, v^p)$ for the tangent space $T(S^p)_y$. For each $x \, \varepsilon \, f^{-1}(y)$ recall from page 12 that

$$df_x : TM_x \to T(S^p)_y$$

maps the subspace $Tf^{-1}(y)_x$ to zero and maps its orthogonal complement $Tf^{-1}(y)_x^\perp$ isomorphically onto $T(S^p)_y$. Hence there is a unique vector

$$w^i(x) \, \varepsilon \, Tf^{-1}(y)_x^\perp \subset TM_x$$

that maps into v^i under df_x. It will be convenient to use the notation $\mathfrak{w} = f^*\mathfrak{v}$ for the resulting framing $w^1(x), \cdots, w^p(x)$ of $f^{-1}(y)$.

DEFINITION. This framed manifold $(f^{-1}(y), f^*\mathfrak{v})$ will be called the *Pontryagin manifold* associated with f.

Of course f has many Pontryagin manifolds, corresponding to different choices of y and $\mathfrak{v}$, but they all belong to a single framed cobordism class:

Theorem A. *If y' is another regular value of f and $\mathfrak{v}'$ is a positively oriented basis for $T(S^p)_{y'}$, then the framed manifold $(f^{-1}(y'), f^*\mathfrak{v}')$ is framed cobordant to $(f^{-1}(y), f^*\mathfrak{v})$.*

Theorem B. *Two mappings from M to S^p are smoothly homotopic if and only if the associated Pontryagin manifolds are framed cobordant.*

44 §7. Framed cobordism

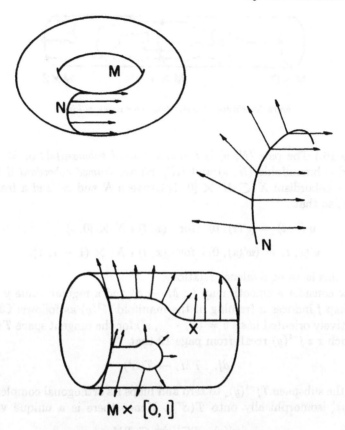

Figure 16. Framed submanifolds and a framed cobordism

Theorem C. *Any compact framed submanifold $(N, \mathfrak{w})$ of codimension p in M occurs as Pontryagin manifold for some smooth mapping $f : M \to S^p$.*

Thus the homotopy classes of maps are in one-one correspondence with the framed cobordism classes of submanifolds.

The proof of Theorem A will be very similar to the arguments in §§4 and 5. It will be based on three lemmas.

Lemma 1. *If $\mathfrak{v}$ and $\mathfrak{v}'$ are two different positively oriented bases at y, then the Pontryagin manifold $(f^{-1}(y), f^*\mathfrak{v})$ is framed cobordant to $(f^{-1}(y), f^*\mathfrak{v}')$.*

PROOF. Choose a smooth path from $\mathfrak{v}$ to $\mathfrak{v}'$ in the space of all positively oriented bases for $T(S^p)_y$. This is possible since this space of bases

The Pontryagin construction 45

can be identified with the space $GL^+(p, R)$ of matrices with positive determinant, and hence is connected. Such a path gives rise to the required framing of the cobordism $f^{-1}(y) \times [0, 1]$.

By abuse of notation we will often delete reference to $f^*\mathfrak{v}$ and speak simply of "the framed manifold $f^{-1}(y)$."

Lemma 2. *If y is a regular value of f, and z is sufficiently close to y, then $f^{-1}(z)$ is framed cobordant to $f^{-1}(y)$.*

PROOF. Since the set $f(C)$ of critical values is compact, we can choose $\epsilon > 0$ so that the ϵ-neighborhood of y contains only regular values. Given z with $\|z - y\| < \epsilon$, choose a smooth one-parameter family of rotations (i.e. an isotopy) $r_t : S^p \to S^p$ so that $r_1(y) = z$, and so that

1) r_t is the identity for $0 \leq t < \epsilon'$,
2) r_t equals r_1 for $1 - \epsilon' < t \leq 1$, and
3) each $r_t^{-1}(z)$ lies on the great circle from y to z, and hence is a regular value of f.

Define the homotopy

$$F : M \times [0, 1] \to S^p$$

by $F(x, t) = r_t f(x)$. For each t note that z is a regular value of the composition

$$r_t \circ f : M \to S^p.$$

It follows a fortiori that z is a regular value for the mapping F. Hence

$$F^{-1}(z) \subset M \times [0, 1]$$

is a framed manifold and provides a framed cobordism between the framed manifolds $f^{-1}(z)$ and $(r_1 \circ f)^{-1}(z) = f^{-1}r_1^{-1}(z) = f^{-1}(y)$. This proves Lemma 2.

Lemma 3. *If f and g are smoothly homotopic and y is a regular value for both, then $f^{-1}(y)$ is framed cobordant to $g^{-1}(y)$.*

PROOF. Choose a homotopy F with

$$F(x, t) = f(x) \quad 0 \leq t < \epsilon,$$
$$F(x, t) = g(x) \quad 1 - \epsilon < t \leq 1.$$

Choose a regular value z for F which is close enough to y so that $f^{-1}(z)$ is framed cobordant to $f^{-1}(y)$ and so that $g^{-1}(z)$ is framed cobordant to $g^{-1}(y)$. Then $F^{-1}(z)$ is a framed manifold and provides a framed cobordism between $f^{-1}(z)$ and $g^{-1}(z)$. This proves Lemma 3.

PROOF OF THEOREM A. Given any two regular values y and z for f, we can choose rotations

$$r_t : S^p \to S^p$$

so that r_0 is the identity and $r_1(y) = z$. Thus f is homotopic to $r_1 \circ f$; hence $f^{-1}(z)$ is framed cobordant to

$$(r_1 \circ f)^{-1}(z) = f^{-1} r_1^{-1}(z) = f^{-1}(y).$$

This completes the proof of Theorem A.

The proof of Theorem C will be based on the following: Let $N \subset M$ be a framed submanifold of codimension p with framing $\mathfrak{v}$. Assume that N is compact and that $\partial N = \partial M = \varnothing$.

Product Neighborhood Theorem. *Some neighborhood of N in M is diffeomorphic to the product $N \times R^p$. Furthermore the diffeomorphism can be chosen so that each $x \in N$ corresponds to $(x, 0) \in N \times R^p$ and so that each normal frame $\mathfrak{v}(x)$ corresponds to the standard basis for R^p.*

REMARK. Product neighborhoods do not exist for arbitrary submanifolds. (Compare Figure 17.)

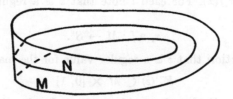

Figure 17. An unframable submanifold

PROOF. First suppose that M is the euclidean space R^{n+p}. Consider the mapping $g : N \times R^p \to M$, defined by

$$g(x; t_1, \cdots, t_p) = x + t_1 v^1(x) + \cdots + t_p v^p(x).$$

Clearly $dg_{(x;0,\cdots,0)}$ is nonsingular; hence g maps some neighborhood of $(x, 0) \in N \times R^p$ diffeomorphically onto an open set.

We will prove that g is one-one on the entire neighborhood $N \times U_\epsilon$ of $N \times 0$, providing that $\epsilon > 0$ is sufficiently small; where U_ϵ denotes the ϵ-neighborhood of 0 in R^p. For otherwise there would exist pairs $(x, u) \neq (x', u')$ in $N \times R^p$ with $\|u\|$ and $\|u'\|$ arbitrarily small and with

$$g(x, u) = g(x', u').$$

The Pontryagin construction

Since N is compact, we could choose a sequence of such pairs with x converging, say to x_0, with x' converging to x_0', and with $u \to 0$ and $u' \to 0$. Then clearly $x_0 = x_0'$, and we have contradicted the statement that g is one-one in a neighborhood of $(x_0, 0)$.

Thus g maps $N \times U_\epsilon$ diffeomorphically onto an open set. But U_ϵ is diffeomorphic to the full euclidean space R^p under the correspondence

$$u \to u/(1 - ||u||^2/\epsilon^2).$$

Since $g(x, 0) = x$, and since $dg_{(x,0)}$ does what is expected of it, this proves the Product Neighborhood Theorem for the special case $M = R^{n+p}$.

For the general case it is necessary to replace straight lines in R^{n+p} by geodesics in M. More precisely let $g(x; t_1, \cdots, t_p)$ be the endpoint of the geodesic segment of length $||t_1 v^1(x) + \cdots + t_p v^p(x)||$ in M which starts at x with the initial velocity vector

$$t_1 v^1(x) + \cdots + t_p v^p(x) / ||t_1 v^1(x) + \cdots + t_p v^p(x)||.$$

The reader who is familiar with geodesics will have no difficulty in checking that

$$g : N \times U_\epsilon \to M$$

is well defined and smooth, for ϵ sufficiently small. The remainder of the proof proceeds exactly as before.

PROOF OF THEOREM C. Let $N \subset M$ be a compact, boundaryless, framed submanifold. Choose a product representation

$$g : N \times R^p \to V \subset M$$

for a neighborhood V of N, as above, and define the projection

$$\pi : V \to R^p$$

by $\pi(g(x, y)) = y$. (See Figure 18.) Clearly 0 is a regular value, and $\pi^{-1}(0)$ is precisely N with its given framing.

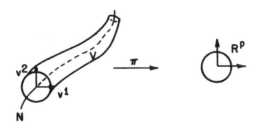

Figure 18. *Constructing a map with given Pontryagin manifold*

§7. Framed cobordism

Now choose a smooth map $\varphi : R^p \to S^p$ which maps every x with $\|x\| \geq 1$ into a base point s_0, and maps the open unit ball in R^p diffeomorphically* onto $S^p - s_0$. Define

$$f : M \to S^p$$

by

$$f(x) = \varphi(\pi(x)) \quad \text{for} \quad x \in V$$

$$f(x) = s_0 \quad \text{for} \quad x \notin V.$$

Clearly f is smooth, and the point $\varphi(0)$ is a regular value of f. Since the corresponding Pontryagin manifold

$$f^{-1}(\varphi(0)) = \pi^{-1}(0)$$

is precisely equal to the framed manifold N, this completes the proof of Theorem C.

In order to prove Theorem B we must first show that the Pontryagin manifold of a map determines its homotopy class. Let $f, g : M \to S^p$ be smooth maps with a common regular value y.

Lemma 4. *If the framed manifold* $(f^{-1}(y), f^*\mathfrak{v})$ *is equal to* $(g^{-1}(y), g^*\mathfrak{v})$, *then f is smoothly homotopic to g.*

PROOF. It will be convenient to set $N = f^{-1}(y)$. The hypothesis that $f^*\mathfrak{v} = g^*\mathfrak{v}$ means that $df_x = dg_x$ for all $x \in N$.

First suppose that f actually coincides with g throughout an entire neighborhood V of N. Let $h : S^p - y \to R^p$ be stereographic projection. Then the homotopy

$$H(x, t) = f(x) \qquad \text{for} \quad x \in V$$

$$H(x, t) = h^{-1}[t \cdot h(f(x)) + (1 - t) \cdot h(g(x))] \quad \text{for} \quad x \in M - N$$

proves that f is smoothly homotopic to g.

Thus is suffices to deform f so that it coincides with g in some small neighborhood of N, being careful not to map any new points into y during the deformation. Choose a product representation

$$N \times R^p \to V \subset M$$

for a neighborhood V of N, where V is small enough so that $f(V)$ and

---

* For example, $\varphi(x) = h^{-1}(x/\lambda(\|x\|^2))$, where h is the stereographic projection from s_0 and where λ is a smooth monotone decreasing function with $\lambda(t) > 0$ for $t < 1$ and $\lambda(t) = 0$ for $t \geq 1$.

The Pontryagin construction

$g(V)$ do not contain the antipode $\bar{y}$ of y. Identifying V with $N \times R^p$ and identifying $S^p - \bar{y}$ with R^p, we obtain corresponding mappings

$$F, G : N \times R^p \to R^p,$$

with

$$F^{-1}(0) = G^{-1}(0) = N \times 0,$$

and with

$$dF_{(x,0)} = dG_{(x,0)} = \text{(projection to } R^p\text{)}$$

for all $x \in N$.

We will first find a constant c so that

$$F(x, u) \cdot u > 0, \quad G(x, u) \cdot u > 0$$

for $x \in N$ and $0 < ||u|| < c$. That is, the points $F(x, u)$ and $G(x, u)$ belong to the same open half-space in R^p. So the homotopy

$$(1 - t)F(x, u) + tG(x, u)$$

between F and G will not map any new points into 0, at least for $||u|| < c$.

By Taylor's theorem

$$||F(x, u) - u|| \leq c_1 ||u||^2, \quad \text{for} \quad ||u|| \leq 1.$$

Hence

$$|(F(x, u) - u) \cdot u| \leq c_1 ||u||^3$$

and

$$F(x, u) \cdot u \geq ||u||^2 - c_1 ||u||^3 > 0$$

for $0 < ||u|| < c = \text{Min } (c_1^{-1}, 1)$, with a similar inequality for G.

To avoid moving distant points we select a smooth map $\lambda : R^p \to R$ with

$$\lambda(u) = 1 \quad \text{for} \quad ||u|| \leq c/2$$

$$\lambda(u) = 0 \quad \text{for} \quad ||u|| \geq c.$$

Now the homotopy

$$F_t(x, u) = [1 - \lambda(u)t]F(x, u) + \lambda(u)tG(x, u)$$

deforms $F = F_0$ into a mapping F_1 that (1) coincides with G in the region $||u|| < c/2$, (2) coincides with F for $||u|| \geq c$, and (3) has no new zeros. Making a corresponding deformation of the original mapping f, this clearly completes the proof of Lemma 4.

PROOF OF THEOREM B. If f and g are smoothly homotopic, then Lemma 3 asserts that the Pontryagin manifolds $f^{-1}(y)$ and $g^{-1}(y)$ are framed cobordant. Conversely, given a framed cobordism $(X, \mathfrak{w})$ between $f^{-1}(y)$ and $g^{-1}(y)$, an argument completely analogous to the proof of Theorem C constructs a homotopy

$$F : M \times [0, 1] \to S^p$$

whose Pontryagin manifold $(F^{-1}(y), F^*\mathfrak{v})$ is precisely equal to $(X, \mathfrak{w})$. Setting $F_t(x) = F(x, t)$, note that the maps F_0 and f have exactly the same Pontryagin manifold. Hence $F_0 \sim f$ by Lemma 4; and similarly $F_1 \sim g$. Therefore $f \sim g$, which completes the proof of Theorem B.

REMARKS. Theorems A, B, and C can easily be generalized so as to apply to a manifold M with boundary. The essential idea is to consider only mappings which carry the boundary into a base point s_0. The homotopy classes of such mappings

$$(M, \partial M) \to (S^p, s_0)$$

are in one-one correspondence with the cobordism classes of framed submanifolds

$$N \subset \text{Interior}(M)$$

of codimension p. If $p \geq \frac{1}{2}m + 1$, then this set of homotopy classes can be given the structure of an abelian group, called the p-th *cohomotopy group* $\pi^p(M, \partial M)$. The composition operation in $\pi^p(M, \partial M)$ corresponds to the union operation for disjoint framed submanifolds of Interior (M). (Compare §8, Problem 17.)

THE HOPF THEOREM

As an example, let M be a connected and oriented manifold of dimension $m = p$. A framed submanifold of codimension p is just a finite set of points with a preferred basis at each. Let $\text{sgn}(x)$ equal $+1$ or -1 according as the preferred basis determines the right or wrong orientation. Then $\sum \text{sgn}(x)$ is clearly equal to the degree of the associated map $M \to S^m$. But it is not difficult to see that the framed cobordism class of the 0-manifold is uniquely determined by this integer $\sum \text{sgn}(x)$. Thus we have proved the following.

Theorem of Hopf. *If M is connected, oriented, and boundaryless, then two maps $M \to S^m$ are smoothly homotopic if and only if they have the same degree.*

On the other hand, suppose that M is not orientable. Then given a basis for TM_x we can slide x around M in a closed loop so as to transform the given basis into one of opposite orientation. An easy argument then proves the following:

Theorem. *If M is connected but nonorientable, then two maps $M \to S^m$ are homotopic if and only if they have the same mod 2 degree.*

The theory of framed cobordism was introduced by Pontryagin in order to study homotopy classes of mappings

$$S^m \to S^p$$

with $m > p$. For example if $m = p + 1 \geq 4$, there are precisely two homotopy classes of mappings $S^m \to S^p$. Pontryagin proved this result by classifying framed 1-manifolds in S^m. With considerably more difficulty he was able to show that there are just two homotopy classes also in the case $m = p + 2 \geq 4$, using framed 2-manifolds. However, for $m - p > 2$ this approach to the problem runs into manifold difficulties.

It has since turned out to be easier to enumerate homotopy classes of mappings by quite different, more algebraic methods.* Pontryagin's construction is, however, a double-edged tool. It not only allows us to translate information about manifolds into homotopy theory; it conversely enables us to translate any information about homotopy into manifold theory. Some of the deepest work in modern topology has come from the interplay of these two theories. René Thom's work on cobordism is an important example of this. (References [36], [21].)

---

* See for example S.-T. Hu, *Homotopy Theory*.

§8. EXERCISES

IN CONCLUSION here are some problems for the reader.

PROBLEM 1. Show that the degree of a composition $g \circ f$ is equal to the product (degree g)(degree f).

PROBLEM 2. Show that every complex polynomial of degree n gives rise to a smooth map from the Gauss sphere S^2 to itself of degree n.

PROBLEM 3. If two maps f and g from X to S^p satisfy $\|f(x) - g(x)\| < 2$ for all x, prove that f is homotopic to g, the homotopy being smooth if f and g are smooth.

PROBLEM 4. If X is compact, show that every continuous map $X \to S^p$ can be uniformly approximated by a smooth map. If two smooth maps $X \to S^p$ are continuously homotopic, show that they are smoothly homotopic.

PROBLEM 5. If $m < p$, show that every map $M^m \to S^p$ is homotopic to a constant.

PROBLEM 6. (Brouwer). Show that any map $S^n \to S^n$ with degree different from $(-1)^{n+1}$ must have a fixed point.

PROBLEM 7. Show that any map $S^n \to S^n$ of odd degree must carry some pair of antipodal points into a pair of antipodal points.

PROBLEM 8. Given smooth manifolds $M \subset R^k$ and $N \subset R^l$, show that the tangent space $T(M \times N)_{(x,y)}$ is equal to $TM_x \times TN_y$.

PROBLEM 9. The graph Γ of a smooth map $f : M \to N$ is defined to be the set of all $(x, y) \in M \times N$ with $f(x) = y$. Show that Γ is a smooth

Exercises

manifold and that the tangent space

$$T\Gamma_{(x,y)} \subset TM_x \times TN_y$$

is equal to the graph of the linear map df_x.

PROBLEM 10. Given $M \subset R^k$, show that the *tangent bundle space*

$$TM = \{(x,v) \, \epsilon \, M \times R^k \mid v \, \epsilon \, TM_x\}$$

is also a smooth manifold. Show that any smooth map $f : M \to N$ gives rise to a smooth map

$$df : TM \to TN$$

where

$$d(\text{identity}) = \text{identity}, \quad d(g \circ f) = (dg) \circ (df).$$

PROBLEM 11. Similarly show that the *normal bundle space*

$$E = \{(x,v) \, \epsilon \, M \times R^k \mid v \perp TM_x\}$$

is a smooth manifold. If M is compact and boundaryless, show that the correspondence

$$(x,v) \mapsto x + v$$

from E to R^k maps the ϵ-neighborhood of $M \times 0$ in E diffeomorphically onto the ϵ-neighborhood N_ϵ of M in R^k. (Compare the Product Neighborhood Theorem in §7.)

PROBLEM 12. Define $r : N_\epsilon \to M$ by $r(x + v) = x$. Show that $r(x + v)$ is closer to $x + v$ than any other point of M. Using this retraction r, prove the analogue of Problem 4 in which the sphere S^p is replaced oy a manifold M.

PROBLEM 13. Given disjoint manifolds $M, N \subset R^{k+1}$, the *linking map*

$$\lambda : M \times N \to S^k$$

is defined by $\lambda(x, y) = (x - y)/\|x - y\|$. If M and N are compact, oriented, and boundaryless, with total dimension $m + n = k$, then the degree of λ is called the *linking number* $l(M, N)$. Prove that

$$l(N, M) = (-1)^{(m+1)(n+1)} l(M, N).$$

If M bounds an oriented manifold X disjoint from N, prove that $l(M, N) = 0$. Define the linking number for disjoint manifolds in the sphere S^{m+n+1}

§8. *Exercises*

Problem 14, The Hopf Invariant. If $y \neq z$ are regular values for a map $f : S^{2p-1} \to S^p$, then the manifolds $f^{-1}(y)$, $f^{-1}(z)$ can be oriented as in §5; hence the linking number $l(f^{-1}(y), f^{-1}(z))$ is defined.

a) Prove that this linking number is locally constant as a function of y.

b) If y and z are regular values of g also, where

$$||f(x) - g(x)|| < ||y - z||$$

for all x, prove that

$$l(f^{-1}(y), f^{-1}(z)) = l(g^{-1}(y), f^{-1}(z)) = l(g^{-1}(y), g^{-1}(z)).$$

c) Prove that $l(f^{-1}(y), f^{-1}(z))$ depends only on the homotopy class of f, and does not depend on the choice of y and z.

This integer $H(f) = l(f^{-1}(y), f^{-1}(z))$ is called the *Hopf invariant* of f. (Reference [15].)

Problem 15. If the dimension p is odd, prove that $H(f) = 0$. For a composition

$$S^{2p-1} \xrightarrow{f} S^p \xrightarrow{g} S^p$$

prove that $H(g \circ f)$ is equal to $H(f)$ multiplied by the square of the degree of g.

The *Hopf fibration* $\pi : S^3 \to S^2$ is defined by

$$\pi(x_1, x_2, x_3, x_4) = h^{-1}((x_1 + ix_2)/(x_3 + ix_4))$$

where h denotes stereographic projection to the complex plane. Prove that $H(\pi) = 1$.

Problem 16. Two submanifolds N and N' of M are said to *intersect transversally* if, for each $x \in N \cap N'$, the subspaces TN_x and TN'_x together generate TM_x. (If $n + n' < m$ this means that $N \cap N' = \emptyset$.) If N is a framed submanifold, prove that it can be deformed slightly so as to intersect a given N' transversally. Prove that the resulting intersection is a smooth manifold.

Problem 17. Let $\Pi^p(M)$ denote the set of all framed cobordism classes of codimension p in M. Use the transverse intersection operation to define a correspondence

$$\Pi^p(M) \times \Pi^q(M) \to \Pi^{p+q}(M).$$

If $p \geq \frac{1}{2}m + 1$, use the disjoint union operation to make $\Pi^p(M)$ into an abelian group. (Compare p. 50.)

APPENDIX
CLASSIFYING 1-MANIFOLDS

WE WILL prove the following result, which has been assumed in the text. A brief discussion of the classification problem for higher dimensional manifolds will also be given.

Theorem. *Any smooth, connected 1-dimensional manifold is diffeomorphic either to the circle S^1 or to some interval of real numbers.*

(An *interval* is a connected subset of R which is not a point. It may be finite or infinite; closed, open, or half-open.)

Since any interval is diffeomorphic* either to [0, 1], (0, 1], or (0, 1), it follows that there are only four distinct connected 1-manifolds.

The proof will make use of the concept of "arc-length." Let I be an interval.

DEFINITION. A map $f : I \to M$ is a *parametrization by arc-length* if f maps I diffeomorphically onto an open subset† of M, and if the "velocity vector" $df_s(1) \in TM_{f(s)}$ has unit length, for each $s \in I$.

Any given local parametrization $I' \to M$ can be transformed into a parametrization by arc-length by a straightforward change of variables.

Lemma. *Let $f : I \to M$ and $g : J \to M$ be parametrizations by arc-length. Then $f(I) \cap g(J)$ has at most two components. If it has only one component, then f can be extended to a parametrization by arc-length of the union $f(I) \cup g(J)$. If it has two components, then M must be diffeomorphic to S^1.*

---
* For example, use a diffeomorphism of the form
$$f(t) = a \tanh(t) + b.$$
† Thus I can have boundary points only if M has boundary points.

PROOF. Clearly $g^{-1} \circ f$ maps some relatively open subset of I diffeomorphically onto a relatively open subset of J. Furthermore the derivative of $g^{-1} \circ f$ is equal to ± 1 everywhere.

Consider the graph $\Gamma \subset I \times J$, consisting of all (s, t) with $f(s) = g(t)$. Then Γ is a closed subset of $I \times J$ made up of line segments of slope ± 1. Since Γ is closed and $g^{-1} \circ f$ is locally a diffeomorphism, these line segments cannot end in the interior of $I \times J$, but must extend to the boundary. Since $g^{-1} \circ f$ is one-one and single valued, there can be at most one of these segments ending on each of the four edges of the rectangle $I \times J$. Hence Γ has at most two components. (See Figure 19.) Furthermore, if there are two components, the two must have the same slope.

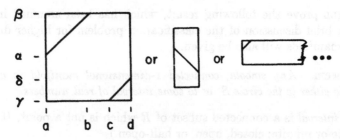

Figure 19. Three of the possibilities for Γ

If Γ is connected, then $g^{-1} \circ f$ extends to a linear map $L : R \to R$. Now f and $g \circ L$ piece together to yield the required extension

$$F : I \cup L^{-1}(J) \to f(I) \cup g(J).$$

If Γ has two components, with slope say $+1$, they must be arranged as in the left-hand rectangle of Figure 19. Translating the interval $J = (\gamma, \beta)$ if necessary, we may assume that $\gamma = c$ and $\delta = d$, so that

$$a < b \leq c < d \leq \alpha < \beta.$$

Now setting $\theta = 2\pi t/(\alpha - a)$, the required diffeomorphism

$$h : S^1 \to M$$

is defined by the formula

$$h(\cos \theta, \sin \theta) = f(t) \quad \text{for} \quad a < t < d,$$
$$= g(t) \quad \text{for} \quad c < t < \beta.$$

The image $h(S^1)$, being compact and open in M, must be the entire manifold M. This proves the lemma.

PROOF OF CLASSIFICATION THEOREM. Any parametrization by arc-length can be extended to one

$$f : I \to M$$

which is maximal in the sense that f cannot be extended over any larger interval as a parametrization by arc-length: it is only necessary to extend f as far as possible to the left and then as far as possible to the right.

If M is not diffeomorphic to S^1, we will prove that f is onto, and hence is a diffeomorphism. For if the open set $f(I)$ were not all of M, there would be a limit point x of $f(I)$ in $M - f(I)$. Parametrizing a neighborhood of x by arc-length and applying the lemma, we would see that f can be extended over a larger interval. This contradicts the assumption that f is maximal and hence completes the proof.

REMARKS. For manifolds of higher dimension the classification problem becomes quite formidable. For 2-dimensional manifolds, a thorough exposition has been given by Kerékjártó [17]. The study of 3-dimensional manifolds is very much a topic of current research. (See Papakyriakopoulos [26].) For compact manifolds of dimension ≥ 4 the classification problem is actually unsolvable.* But for high dimensional simply connected manifolds there has been much progress in recent years, as exemplified by the work of Smale [31] and Wall [37].

---

* See Markov [19].

BIBLIOGRAPHY

THE following is a miscellaneous list consisting of original sources and of recommended textbooks. For the reader who wishes to pursue the study of differential topology, let me recommend Milnor [22], Munkres [25], and Pontryagin [28]. The survey articles [23] and [32] should also prove useful. For background knowledge in closely related fields, let me recommend Hilton and Wylie [11], Hu [16], Lang [18], de Rham [29], Steenrod [34], and Sternberg [35].

[1] Allendoerfer, C. B., "The Euler number of a Riemann manifold," *Amer. Jour. Math.* 62 (1940), 243-248.
[2] Apostol, T. M., *Mathematical Analysis*. Reading, Mass.: Addison-Wesley, 1957.
[3] Auslander, L., and R. MacKenzie, *Introduction to Differentiable Manifolds*. New York: McGraw-Hill, 1963.
[4] Brouwer, L. E. J., "Uber Abbildung von Mannigfaltigkeiten", *Math. Annalen* 71 (1912), 97-115.
[5] Brown, A. B., "Functional dependence," *Trans. Amer. Math. Soc.* 38 (1935), 379-394. (See Theorem 3-III.)
[6] Chern, S. S., "A simple intrinsic proof of the Gauss-Bonnet formula for closed Riemannian manifolds," *Annals of Math.* 45 (1944), 747-752.
[7] Dieudonné, J., *Foundations of Modern Analysis*. New York: Academic Press, 1960.
[8] Dubovickiĭ, A. Ya., "On differentiable mappings of an n-dimensional cube into a k-dimensional cube," *Mat. Sbornik* N.S. 32 (74), (1953), 443-464. (In Russian.)
[9] Fenchel, W., "On total curvatures of Riemannian manifolds," *Jour. London Math. Soc.* 15 (1940), 15-22.
[10] Goffman, C., *Calculus of Several Variables*. New York: Harper & Row, 1965.
[11] Hilton, P., and S. Wylie, *Homology Theory*. Cambridge Univ. Press, 1960.

[12] Hirsch, M., "A proof of the nonretractibility of a cell onto its boundary," *Proc. Amer. Math. Soc.* 14 (1963), 364–365.
[13] Hopf, H., "Abbildungsklassen n-dimensionaler Mannigfaltigkeiten," *Math. Annalen* 96 (1926), 209–224.
[14] ———, "Vektorfelder in n-dimensionalen Mannigfaltigkeiten," *Math. Annalen* 96 (1926), 225–250.
[15] ———, "Uber die Abbildungen von Sphären auf Sphären niedrigerer Dimension," *Fundamenta Mathematicae* 25 (1935), 427–440.
[16] Hu, S.-T., *Homotopy Theory*. New York: Academic Press, 1959.
[17] Kerékjártó, B. v., *Vorlesungen über Topologie*. Berlin: Springer, 1923.
[18] Lang, S., *Introduction to Differentiable Manifolds*. New York: Interscience, 1962.
[19] Markov, A. A., "Insolubility of the problem of homeomorphy," *Proceedings Intern. Congress of Math. 1958*, Cambridge Univ. Press, 1960, pp. 300–306. (In Russian.)
[20] Milnor, J., "On manifolds homeomorphic to the 7-sphere," *Annals of Math.* 64 (1956), 399–405.
[21] ———, "A survey of cobordism theory," *L'Enseignement math.* 8 (1962), 16–23.
[22] ———, *Morse Theory*. (Annals Studies 51.) Princeton Univ. Press, 1963.
[23] ———, "Differential topology," *Lectures on Modern Mathematics*, II, ed. T. L. Saaty, New York: Wiley, 1964, pp. 165–183.
[24] Morse, A. P., "The behavior of a function on its critical set," *Annals of Math.* 40 (1939), 62–70.
[25] Munkres, J. R., *Elementary Differential Topology*. (Annals Studies 54). Princeton Univ. Press, 1963.
[26] Papakyriakopoulos, C. D., "The theory of three-dimensional manifolds since 1950," *Proceedings Intern. Congress of Math. 1958*, Cambridge Univ. Press, 1960, pp. 433–440.
[27] Pontryagin, L. S., "A classification of continuous transformations of a complex into a sphere," *Doklady Akad. Nauk. S.S.S.R. (Comptes Rendues)* 19 (1938), 147–149.
[28] ———, "Smooth manifolds and their applications in homotopy theory," *Amer. Math. Soc. Translations*, Ser. 2, II (1959), 1–114. (Translated from *Trudy Inst. Steklov* 45 (1955).)
[29] Rham, G. de, *Variétés différentiables*. Paris: Hermann, 1955.
[30] Sard, A., "The measure of the critical points of differentiable maps," *Bull. Amer. Math. Soc.* 48 (1942), 883–890.
[31] Smale, S., "Generalized Poincaré's conjecture in dimensions greater than four," *Annals of Math.* 74 (1961), 391–406.
[32] ———, "A survey of some recent developments in differential topology," *Bull. Amer. Math. Soc.* 69 (1963), 131–145.
[33] Spivak, M., *Calculus on Manifolds*. New York: Benjamin, 1965.
[34] Steenrod, N., *The Topology of Fibre Bundles*. Princeton Univ. Press, 1951.

[35] Sternberg, S., *Lectures on Differential Geometry*. New York: Prentice-Hall, 1964.
[36] Thom, R., "Quelques propriétés globales des variétés différentiables," *Commentarii Math. Helvet.* 28 (1954), 17–86.
[37] Wall, C. T. C., "Classification of $(n-1)$-connected $2n$-manifolds," *Annals of Math.* 75 (1962), 163–189.
[38] Whitney, H., "A function not constant on a connected set of critical points," *Duke Math. Jour.* 1 (1935), 514–517.

Added April, 1969:
[39] Husemoller, D., *Fiber Bundles*. New York: McGraw-Hill, 1966.
[40] Spanier, E. H., *Algebraic Topology*. New York: McGraw-Hill, 1966.
[41] Wall, C. T. C., "Topology of smooth manifolds," *J. London Math. Soc.* 40 (1965), 1–20.
[42] Wallace, A. H., *Differential Topology, First Steps*. New York: Benjamin, 1968.

PAGE-BARBOUR LECTURE SERIES

The Page-Barbour Lecture Foundation was founded in 1907 by a gift from Mrs. Thomas Nelson Page (née Barbour) and the Honorable Thomas Nelson Page for the purpose of bringing to the University of Virginia each session a series of lectures by an eminent person in some field of scholarly endeavor. The materials in this volume were presented by Professor John W. Milnor in December, 1963, as the forty-seventh series sponsored by the Foundation.

INDEX

antipodal map, 30, 52
boundary, 12
Brouwer, L. E. J., 35, 52
    degree, 20, 28
    fixed point theorem, 14
Brown, A. B., 10, 11
chain rule for derivatives, 3, 7
cobordism, 42, 43, 51
codimension, 42
cohomotopy, 50
coordinate system, 1
corresponding vector fields, 32
critical point, 8, 11
critical value, 8, 11
degree of a map, 28
    mod two, 20
derivative of a map, 2-7
diffeomorphism, 1
differential topology, 1, 59
dimension of a manifold, 1, 5, 7
disk, 13, 14
Euler number, 35, 36, 40
framed cobordism, 43
framed submanifold, 42, 44, 46
Fubini theorem, 17
fundamental theorem of algebra, 8
Gauss-Bonnet theorem, 38
Gauss mapping, 35, 38
half-space, 12
Hirsch, M., 14

homotopy, 20, 21, 52
Hopf, H., 31, 35, 36, 40, 51, 54
index (of a zero of a vector field), 32-34
index sum, 35-41
inverse function theorem, 4, 8
inward vector, 26
isotopy, 21, 22, 34
linking, 53
Morse, A. P., 10
Morse, M., 40
nondegenerate zero (of a vector field) 37, 40
normal bundle, 53
normal vectors, 11, 12, 42
orientation, 26
    of a boundary, 27
    of $F^{-1}(y)$, 28
outward vector, 26
parametrization, 1, 2
    by arc-length, 55
Poincaré, H., 35
Pontryagin, L., 16, 42
Pontryagin manifold, 43
product neighborhood theorem, 46
regular point, 7
regular value, 8, 11, 13, 14, 20, 27, 40, 43
Sard, A., 10, 16
smooth manifolds, 1
    with boundary, 12

smooth manifolds (cont.)
  the classification problem, 57
  of dimension zero, 2
  of dimension one, 14, 55
  oriented, 26
smooth maps (= smooth mappings),1
sphere, 2

stereographic projection, 9, 48
tangent bundle, 53
tangent space, 2–5
  at a boundary point, 12, 26
tangent vector, 2
vector fields, 30, 32–41
Weierstrass approximation theorem, 14

INDEX OF SYMBOLS

$\deg(f; y)$, 27
df_x, 2–7
D^m, 14
∂X, 12
f^*v, 43
$\#f^{-1}(y)$, 8

$g \circ f$, 1
H^m, 12
R^k, 1
S^{n-1}, 2
TM_x, 2–5
$\|x\|$, 14

Printed and bound by CPI Group (UK) Ltd, Croydon, CR0 4YY
09/06/2025
14685662-0001